Detlef Gysau

Fillers for Paints

Fundamentals and Applications

3rd Revised Edition

VINCENTZ
Providing Professional Information

Cover: Tiberius Gracchus/Fotolia

Detlef Gysau
Fillers for Paints, Fundamentals and Applications, 3rd Revised Edition
Hanover: Vincentz Network, 2017
European Coatings Library
ISBN 978-3-86630-284-6

Vincentz Network, Plathnerstr. 4c, 30175 Hanover, Germany
T +49 511 9910-033, F +49 511 9910-029, books@european-coatings.com

Layout: Schubert, Hamburg, Germany

European Coatings Library

Detlef Gysau

Fillers for Paints

Fundamentals and Applications

3rd Revised Edition

For Jacqueline and Gian-Flurin and Mica-Ladina and also Rambo and Fuchur

There is no debt more pressing than the expression of gratitude.

Marcus Tullius Cicero

Foreword

The topic of fillers for use in paints and varnishes is an old one, so one might ask why there has been no comprehensive book on the subject to date. Could it be something to do with the earlier prevailing perception of fillers as cheap materials for bulking up profits? Are fillers even worth writing about? Certainly! The sheer number of mineral end-products, the frequently underestimated effort that goes into their manufacture, the testing done to characterise their diverse properties, their wide-ranging applications – that is an awful lot of information to pack into a single work without diluting its focus.

Simply to consider the spectrum of professions involved in producing and using fillers – geologists, mineralogists, mechanical engineers, machine operators, chemists, paint and varnish specialists – highlights the extent of hidden technical activity. Fillers are instrumental in many properties of coating materials and films: their rheology, content of volatile organic compounds, solids content, brightness, opacity, reflectivity, adhesion, anti-corrosion characteristics, mechanical and chemical resistance... the list goes on. The bottom line is, proper use of fillers calls for a great deal of knowledge.

The present book sets out to convey that knowledge in a straightforward and understandable manner, without compromising scientific objectivity and rigour. Special attention has been given to clear topical division and structuring, to facilitate finding pertinent information, fast. That having been said, the gamut of available fillers is so vast that there would be insufficient space to cover all the materials out there, some of them quite exotic. Instead, this book concentrates on fillers in regular current use, with numerous figures and tables to illustrate their properties and applications. All the same, this book cannot claim to be exhaustive in scope. Readers wishing to obtain further information and details will be served by the extensive bibliographic references provided.

This book is intended for anyone who is in any way professionally involved with fillers used in coating materials. Beginners and students will gain a comprehensive overview of the field, while experienced developers will find practical details of immediate relevance to solving their everyday problems.

In 2016 I was notified that also the second edition of "Fillers for Paints" is going to be sold out soon as well. I am more than delighted to learn that also the second edition found so many new readers. The continued interest in my book is also judged by manifold feedbacks which I received since 2006. All of them expressed to me their thanks and congratulations by filling a knowledge gap in raw materials for paints.

In particular, I appreciate that the book supports training for all different kind of groups, either in industry or science. The third edition allowed me to place small corrections, up-date market and filler data and add more sub chapters about new fillers and nevertheless an outlook about the future, for example sustainability and light weight fillers.

Detlef Gysau
Oftringen/Switzerland, January 2017

Contents

1 Introduction

1.1 Historical overview

Paints and varnishes have a history that goes back around 100,000 years, to the time when stone age peoples applied red body-paint as part of their cultish rituals [1]. The first paintings on cave walls date back to the late Stone Age, their origins still somewhat shrouded in mystery. Many thousands of years later, in the 4th century B.C., the intermingling of ancient Egyptian and Greek civilisations brought remarkable developmental advances through "Hagia Techné" or "Alchimia" – hallowed arts practiced by the high priests of the day. Their discoveries about the secrets of paint making remained influential well beyond the 16th century A.D. As the industrial revolution started in the 18th century, paints and varnishes came into widespread use for many different applications. Early 20th century triumphs of chemistry and technology in particular signified a clear departure from empiricism, to science.

The history of fillers can be traced back almost as far as paints and varnishes. Pigment analysis has revealed the presence of filler materials in early cave paintings [2, 3], the oldest identifiable specimens dating from 20,000 to 30,000 years ago, see Table 1.1 p.14.

However, the first people to systematically use fillers for their cave paintings were the ancient Egyptians, and the Mediterranean cultures that succeeded them. The most frequent materials were chalk and gypsum, both white mineral fillers. Clays, or crushed mollusc shells, were also used on occasion. As history progressed, the ancient Greeks began using a mineral that was whiter still: white lead. Because of its rare occurrence in nature, they developed an intricate process to obtain the pigment synthetically. Contemporary demand for greater opacity and brightness evidently made the effort worthwhile. The Roman historians Pliny and Vitruvius respectively reported eight and five white pigments then in use, although only three were of real significance: the minerals melinum, paraetonium and cerussa (white lead).

During the period of the Roman Empire, there was a marked increase in the consumption of fillers, which were used in paints for murals, panels and frescoes. But filler production collapsed along with the Roman Empire, and artists subsequently resorted to local minerals. There were large chalk deposits in England, France, the Netherlands and Germany. Even in Spain, chalk grew prevalent under the name of Spanish white. In Italy, though, gypsum predominated. That was the situation until the 19th century, when the industrial revolution came into full swing.

The enormous increase in consumption of raw materials during the industrial revolution also brought a sustained rise in demand for fillers. Semi- and fully-automatic dress-

Table 1.1: Natural and synthetic white minerals used by the ancients

Ancient name	Modern name	Mineral composition
Cerussa [1,2]	White lead, Krems White, etc.	Basic lead carbonate, made from metallic lead and vinegar
Creta anularia [1,2]	Crete white	Chalk mixed with powdered glass
Cimolia creta [1]	Kimolos chalk	Chalk or clay-like material
Creta eretria [1]		Probably a white talc, named after a place on the southwest coast of Euboa
Creta selinusia [1,2]	Selinus chalk	Chalk or chalk clay, named after a place on Sicily
Melinum [1,2]	Melian white	Bianca San Giovannini or white clay
Paraetonium [1,2]	White sepiolite	Limestone chalk with some magnesium phosphate, silicic acid and clay, named after a place in Libya
Creta argentaria [1]	Argentiferous chalk	Chalk

[1] Pliny, Natural History, XXXV [2] Vitruvius, Ten Books on Architecture, VII

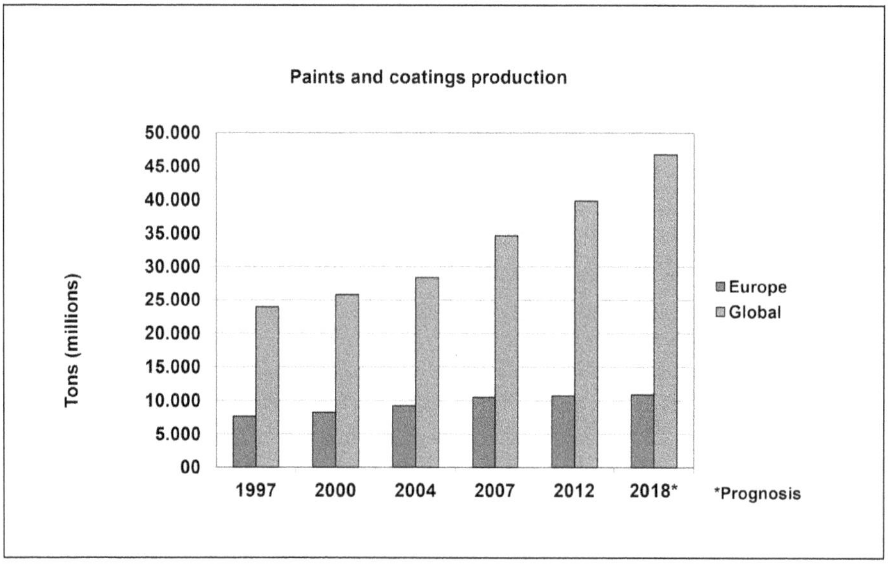

Figure 1.1: Overall European production of paints and varnishes, in millions of tons

source: The Chemical Economics Handbook – SRI Consulting 2009; World Paint & Coatings Industry Association (WPCIA) – Annual Figures ing processes were developed to address this demand, as well as to meet the steadily advancing requirements of industry. High-power machinery like crushers, grinding drums and classifiers came into use. The end of the Second World War brought even greater demand for fillers, which was a motor for further modernisation by the filler industry. The resulting technical developments led to ever-finer natural fillers and tailor-made synthetic fillers, some with surface coatings, see Figure 1.1 p.14.

1.2 Filler market

The market often underrates fillers, on account of their relatively low price compared to the other raw materials used for making paints and varnishes. Overall Global and European show a continuous growth since 1997. Once believing the prognosis for the global paint and coatings market, then the number will almost double from 1997 to 2018 to approx. 47 Mio tons. Despite the growth for the production in Europe, its global share drops from 32.0 % in 1997 versus a much stronger growth in emerging markets such as Asia to 23 % in 2018.

If one compares the four million-plus tons of fillers consumed in 2003 with the quantity of paints and varnishes produced in that year, their 42 percent statistical share makes it clear that fillers are the dominant class of raw materials used in paint and varnish production, see Figure 1.2 p.15.

The chart of mineral fillers in current use reveals another dominance: natural calcium carbonate is the basis for three quarters of all the fillers used in paints and varnishes. Carbonate fillers together have an 85 percent aggregate share. This profile of mineral filler consumption is essentially repeated on other continents as well, see Figure 1.3, p. 16.

An analysis of application areas reveals that most fillers go into architectural paints, in particular emulsion paints. This group of paint systems is far and away the largest, at around 60 percent of overall paint and varnish production. Empirically speaking, classical and contemporary coating systems both tend to use considerably less fillers, or indeed dispense with them altogether. These systems generally are formulated below the critical pigment volume concentration (CPVC), which necessitate a higher proportion of pigments in order to achieve sufficient opacity.

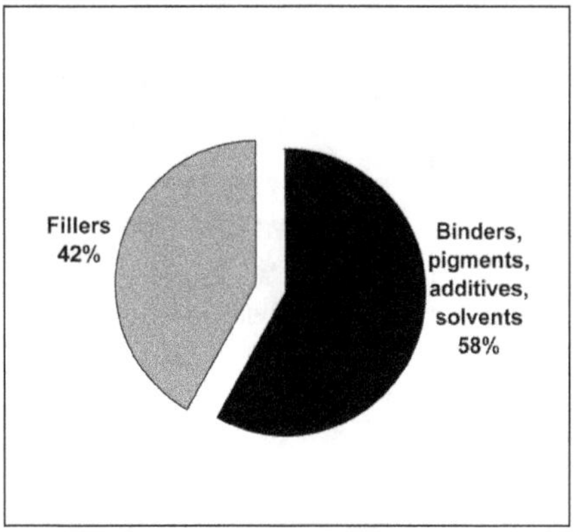

Figure 1.2: Statistical share of fillers in the overall composition of European paints and varnishes

1.3 Definition of fillers and pigments

There are numerous differences in the properties of fillers and pigments. Yet they can also overlap, depending on the application. Therefore, it is important to draw a clear distinction between these two groups of raw materials. Help is provided by the sets of standard specifications published by the German standards institution (DIN) [4, 5], the European Committee for Standardisation (CEN) and the International Organisation for Standardisation (ISO) [6].

According to DIN 55943, EN 971-1 and ISO 3262 part 1, "a filler is a substance consisting of particles which is practically insoluble in the application medium and is used to increase volume or to improve technical properties and/or to influence optical properties." The standards discourage the use of terms like "extender", "extender pigment", or "pigment extender", instead stating that "on this basis, whether a substance should be regarded as a filler or a pigment is determined by its application."

Pigments are defined in the German standards DIN 55943 and DIN 55945: "A pigment is a substance consisting of particles which is practically insoluble in the application medium and is used as a colorant or by virtue of its corrosion-inhibiting or magnetic properties." Pigments may be described more precisely, for example as inorganic or organic pigments, coloured pigments, white pigments, effect pigments, anti-corrosion pigments, magnetic pigments, etc. depending on their chemical composition, optical or other technical properties.

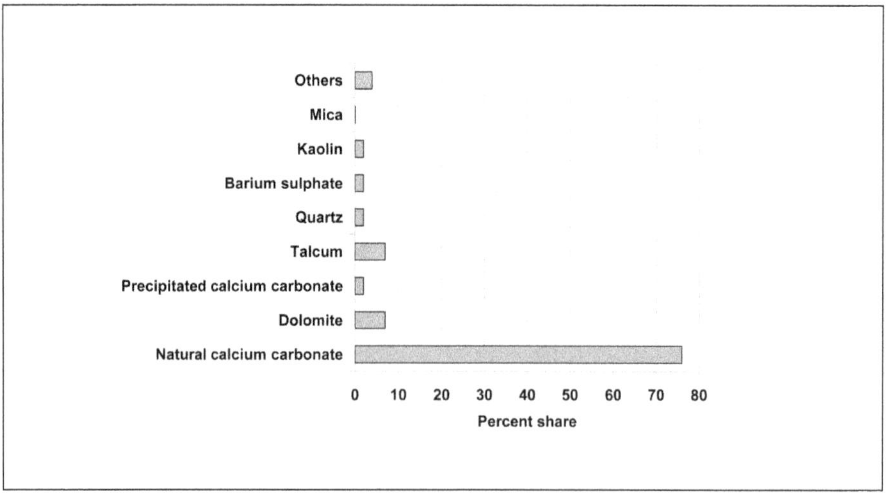

Figure 1.3: Percent share of fillers in Europe, categorised by mineral

Table 1.2: Summary of filler categories and materials for coating applications

Filler primary category	Natural fillers	Synthetic fillers
Carbonates	Natural calcium carbonate (GCC) Dolomite Barium carbonate	Precipitated calcium carbonate (PCC) Modified calcium carbonate (MCC)
Silicas (silicon dioxides)	Quartz Cristobalite Kieselguhr Diatomaceous earth Glass powder Ground pumice	Pyrogenous silicic acid Precipitated silicic acid
Silicates	Talcum Pyrophyllite Chlorite Kaolin Mica Feldspar Wollastonite Ground shale Amphibole Perlite	Precipitated aluminium silicate Precipitated calcium silicate Precipitated calcium-aluminium silicate Precipitated sodium-aluminium silicate
Sulphates	Natural barium sulphate Calcium sulphate	Precipitated barium sulphate
Oxides		Aluminium hydroxide Magnesium hydroxide
Organic fillers	Wood flour Cork flour	Micro hollow spheres Cellulose fibres

They are categorised according to DIN 55944. Older terms like "colouring body", "lake dye", "earth pigment" and "mineral pigment" should no longer be used.

Practically speaking, material constants like the refractive index often determine whether a substance is acting as a pigment or a filler. This is usually apparent from the optical effect of the substance as a component of the coating material. If the substance helps to increase opacity, then it has the characteristics of a pigment. If it behaves transparently, though, it is considered to be a filler. In general, materials with a high refractive index (≥ 1.7) are pigments. All other mineral materials with a similar refractive index, like organic polymers, belong in the category of fillers.

1.4 Classification of fillers

Given the diversity of mineral fillers, it is helpful to divide them into various categories. Categories like carbonates, silicates, silicas (silicon dioxides), sulphates, oxides and organic fillers include well known as well as more obscure materials. In addition to this type of categorisation, fillers are also grouped according to their natural versus synthetic origin.

Not all of the fillers listed in Table 1.2 are (as yet) industrially significant. Although they have been listed here for the sake of completeness, they will not be covered in subsequent chapters of this book.

1.5 References

[1] Pietsch, E., Altamira und die Urgeschichte der chemischen Technologie, Deutsches Museum Abhandlungen und Berichte 31, booklet 1, p. 15, 1963

[2] Science News 125, 348, 1984

[3] Tegethoff, F. W., Calcium Carbonate – From the Cretaceous Period into the 21st Century, p. 55 ff, Birkhäuser Verlag, Basel, 2001

[4] DIN-Taschenbuch 49, Farbmittel, 1. Pigmente, Füllstoffe, Farbstoffe, DIN 5033-1 to DIN 55929: Normen, Beuth, Berlin, 1993

[5] DIN-Taschenbuch 157, Farbmittel, 2. Pigmente, Füllstoffe, Farbstoffe, DIN 55943 to DIN 66131: Normen, Beuth, Berlin, 1993

[6] ISO Standards Handbook, Paints and varnishes, Vol. 3 – Raw materials, ISO 150 to ISO 14900, International Organisation for Standardisation, Geneva, 2002

2 Mineralogy

Current estimates put the age of planet Earth at 4.55 to 4.66 billion years. Industrial minerals are extracted from the Earth's continental crust, which exhibits a complex, planet-wide pattern of belts and mountains with deformed rock series. Somewhat younger than the planet itself are the Earth's minerals. This chapter discusses them in a competent, concise and understandable way.

2.1 Carbonates

Constantly changing continental surfaces, and the rate of the oceans' spread, all bore directly on the sedimentation rate of carbonates. Calcium and magnesium were increasingly exchanged between the oceanic crust and seawater as new areas of seabed emerged around the oceanic ridges. Tectonic movement, metamorphosis and volcanic activity also aided the process, making the Earth's atmosphere progressively warmer and richer in carbon dioxide. This climatic change accelerated the carbonate weathering process, releasing calcium. Tectonic movement also created shallow seas on the continents, which in turn encouraged carbonate sedimentation. Conditions for the formation of carbonate rocks were optimal during the Cambrian period, from the late Devonian to the early Carboniferous, the Permian, Triassic, Jurassic and Cretaceous periods. The largest carbonate formations date from these eras in geological time.

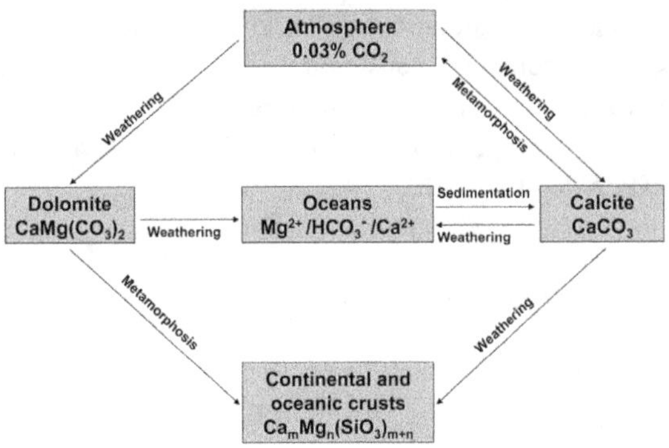

Figure 2.1: The carbonate cycle

Table 2.1: Physical properties and identification numbers of principal carbonates used by the filler industry

Mineral	**Chemical formula**	**CAS no.**	**Density [g/cm³]**	**Refractive index**	**Mohs hardness**
Natural calcium carbonate	$CaCO_3$	1317-65-3	2.7 to 2.9	1.59	3 to 4
Dolomite	$CaMg(CO_3)_2$	16389-88-1	2.85	1.62	3.5 to 4

The carbonates in principal use as industrial fillers are calcium carbonate and magnesium carbonate. Other carbonates and hydrogen carbonates also exist in association with potassium, sodium and ammonium cations. They are generally used by the food industry as additives, and will not be discussed further here.

2.1.1 Calcium carbonate

Calcium carbonate[1–5] has the chemical formula $CaCO_3$. It is the primary constituent of limestone, which is formed by sedimentation of particulate matter. In fact, limestone can be formed in three distinct ways: by chemical precipitation (usually in fresh water), by a biochemical process, or by organogenic sedimentation. The latter two occur in the oceans, due to the high concentration of salt in the water. Various crystal modifications may arise, depending on the kind of organisms involved and the water temperature. The solubility of calcium carbonate in water depends very much on its temperature, and the amount of dissolved carbon dioxide present. Warm ocean water contains little dissolved calcium carbonate: higher temperatures cause the carbonate to precipitate out of solution, eventually forming reefs. In northern latitudes as well as in deep waters, ocean temperatures are low, so the carbon dioxide content is considerably higher. That means the concentration of dissolved calcium carbonate can rise by a hundredfold and more. Chalk shells dissolve entirely beyond a certain depth, 3.5 to 5 km depending on latitude. This is known as the carbonate compensation depth, or CCD.

$$CaCO_3 + CO_2 + H_2O \Leftrightarrow Ca(HCO_3)_2$$

Equation 2.1: Equilibrium reaction for solubility of calcium carbonate in water

Natural calcium carbonate is largely formed by a process of organogenic sedimentation, also known as bioclastic sedimentation. Here, the inorganic remains of invertebrate species like molluscs, corals, sponges, ectoprocta, echinoderms, foraminiferida and algae form deposits on the seabed. The deposited particles can vary greatly in size, depending on the timescale of the deposition process and the degree of compression. The smallest particles are coccoliths measuring just a few micrometers; at the other extreme are whole mollusc shells, several centimetres in size. Pressure causes the sedimentary lime sludge to solidify on ocean floors and form friable sedimentary rock – in other words, chalk. ISO 3262 part 4 defines chalk as a soft, sedimentary rock form of calcium carbonate that results from the chalk formation process. Chalk is typically comprised of microcrystalline calcite crystals with a diameter of around one micrometer. A significant feature that persists even after many millions of years is the mass of shell fragments and skeletons of microscopic marine life forms like coccoliths and foraminifera that chalk contains.

Figure 2.2: Scanning electron microscope photo of ray-like calcite particles constituting the disk-shaped plate of a coccosphere, magnified 6,500 times source: OmyaAG

In warm oceans especially, chalk deposits built up a variety of structures including massive flat reefs up to several hundred metres thick. One such reef limestone belt was formed some 150 million years ago in the late Jurassic period and stretches from the Paris basin all the way to southern Germany. Yet the earliest limestone formations are much older still, with sedimentary deposits dating back as far as the Precambrian era, three billion years ago. Table 2.2 provides a historical perspective.

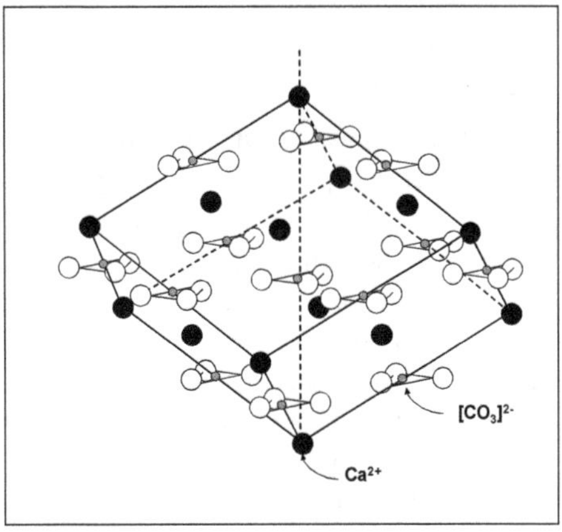

Figure 2.3: Crystalline structure of a single calcite rhombohedron

When a carbonate sediment is formed, it is gradually compacted by the build-up of pressure from layers above, and becomes less porous. The second step is cementation: under increasing pressure, countless pores begin to fill up with precipitated calcite. These two steps are collectively known as diagenesis, a term used to describe the transition from sedimentary sludge to microcrystalline chalk to microcrystalline, dense limestone over many millions of years. Limestone with fine crystalline cement is

Table 2.2: Geological timeline for limestone formations

Era	Age (millions of years)	Period	Significant folding phases	Carbonate sedimentation
Cenozoic	1.6	Quaternary		
	2.3	Tertiary	Alpine chain Pyrenean chain	Limestone
Mesozoic	65	Cretaceous	Laramic phase (North America)	Chalk
	130	Jurassic	Andes chain	Reef limestone, oolitic limestone
	205			
	250	Triassic	Kimeric phase (Asia)	Dolomite
Paleozoic	290	Permian	Dinaric chain (Europe)	
	360	Carboniferous		Dark
		Devonian		limestone
	400	Silurian	Caledonian chain	
	420	Ordovician	(Europe-North	
	500	Cambrian	America)	Earliest
	545			limestone
Precambrian	2600	Proterozoic	Multiple phases	Stromatoliths
	4600	Archean	in ancient plates	

known as micrite, or microcrystalline calcite. Geologists also call coarse crystalline calcite a sparite. Diagenesis can also produce dolomitisation (see Chapter 2.1.2). Deposits of red chalk, sandstone, quartz, phosphate, pyrite and carboniferous chalk are the result of mineral impurities mixed in with the pure limestone. Limestone purity is therefore rated by categories A through D, classified according to ISO 3262 part 5.

The formation of marble requires radical conditions, which can only be achieved under metamorphosis. The transmutation from chalk or limestone to marble takes place under very high pressures in excess of 1,000 bar, coupled with high temperatures between 200 and 500 °C. Limestone melts under these conditions, before returning to the solid state in a very slow recrystallisation process. Transmutation incidentally increases the purity of the calcium carbonate. The resulting fine to coarse-grained metamorphic rock contains over 50 percent by volume of calcite. Perhaps the best-known deposit of extremely white marble is at Carrara, in Italy. The marble mountains date back some 220 million years, a time when the approaching European and African continents caused thick formations from the Tuscan Nappe to slide over an existing limestone deposit. The deposit was thus covered to a depth of five to ten kilometres and subjected to temperatures of around 300 °C over a long period. Recrystallisation ended just 15 million years ago. A final tectonic squeeze caused what are perhaps the finest marble mountains anywhere to bulge back up to the Earth's surface. Not only did metamorphosis cause the limestone to transmute into marble, it also transformed the impurities present into new minerals: wollastonite (a form of calcium silicate), quartz, muscovite and phlogopite (both mica), amphibole, diopside, serpentine, graphite, pyrite, marcasite, chalcopyrite and feldspar, to name some examples.

Calcium carbonate occurs in three distinct crystal modifications, so it is also referred to as a polymorphous, variform compound. The three crystal modifications are: calcite [6], aragonite and vaterite. Calcite, with its trigonal crystal system, is predominant in nature. It also happens to be the most multi-faceted, multi-shaped mineral of all, with a crystal lattice of elementary cells in the form of a rhombohedral prism. In other words, a calcite crystal is delineated by six equivalent (rhombic) surfaces, and appears like a diagonally stretched cube. It was this observation that led to the discovery of the fundamental laws of mineralogy by Abbé Haüy in the 18th century. Calcite crystals occur most frequently as hexagonal prisms and scalenohedrons, whereas the fundamental rhombohedralform is rarer in nature.

Aragonite [7], the scarcer rhombic crystal modification, is named after the northern Spanish region of Aragon. It occurs as a cyclic polycrystal in three characteristic crystalline forms, twins and pseudohexagonal trillings being the most common. Aragonite has a characteristic needle or slab shaped appearance, often with pointed ends.

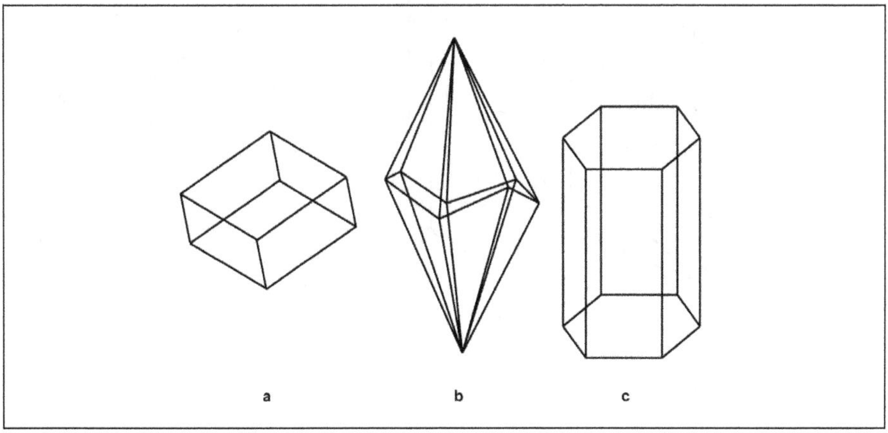

Figure 2.4: Crystalline forms of calcite – a) rhombohedron, b) scalenohedron, c) prism

Vaterite, the third crystal modification of calcium carbonate, is named after the German mineralogist and chemist Heinrich Vater. Its hexagonal crystals are unstable, and hence extremely rare in nature. However, they can be encouraged to form by synthetic precipitation under certain controlled conditions. Vaterite crystals are normally small and fibrous; fine, microscopic platelets are a rarer variant.

As mentioned previously, calcium carbonate is an abundantly occurring mineral. Therefore, only selected examples of well known, industrially viable deposits are listed here. Chalk is found in the Champagne, near Lille and Saint-Omer in France; in the eastern British Isles; Mons in Belgium; Fakse on Jutland in Denmark; near Malmö in Sweden; in Lägerdorf and on the Baltic island of Rügen in Germany; as well as in Poland and Russia. There are limestone deposits in Spain near Zaragossa, Belchite and Granada; on the Greek Ionian islands; at Orgon near Avignon in southern France; in the Friulian Alps and the Abruzzi region of Italy; in Germany's Swabian and Frankish Albs; as well as on Java in Indonesia. Marble deposits exist in Tautavel in the Spanish Pyrenees; Vizarron in Mexico; Kanwon in South Korea; Carrara, Massa, Laas and Sterzing in Italy; Macael in Spain; Estremoz in Portugal; Middlebury and Danby in Vermont, Cristo in California (USA); Gummern and Graz in Austria; Elnesvagen and Bodö-Fauske in Norway; Pargas and Lappeenranta in Finland; in Sweden; and in the Chinese provinces of Liaoning and Jilin.

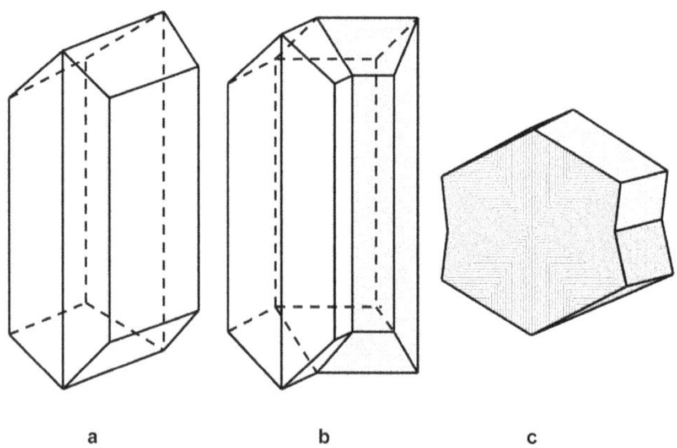

Figure 2.5: Crystalline forms of aragonite – a) monocrystal, b) twin, c) hexagonal trilling

2.1.2 Dolomite

Dolomite[8], named after the French mineralogist Dolomieu, is a carbonatic sedimentary rock that has much in common with calcium carbonate. Chemically, dolomite consists of $CaCO_3$ and $MgCO_3$ in a roughly stoichometric ratio. For this reason, dolomite is occasionally referred to as a double carbonate [$CaMg(CO_3)_2$]. There may be a slightly higher proportion of calcite ($CaCO_3$) or magnesite ($MgCO_3$), depending on the source.

Dolomite forms in a diagenetic environment on chalk-containing sludge or organic layers on salt-water ocean beds. In the dolomitisation process, calcium ions in the crystal lattice are exchanged for magnesium ions in an equilibrium reaction. The equilibrium point depends on the concentration of calcium and magnesium in the solid rock as well as the surrounding environment. Cation exchange may occur during sedimentation in magnesium-rich seawater, or in a subsequent, secondary dolomitisation process as magnesium-containing water filters through previously formed rock. Dedolomitisation can occur in just the same way, although this is very rare. These processes also affect the rock's porosity and density, since the density of dolomite at 2.87 g/cm^3 is higher than that of calcite. Dolomite, with a Mohs hardness of 3.5 to 4, is also a harder material. Like calcite, dolomite has a trigonal crystal system. The rhombohedrons are fully fissionable. The crystal aggregate is dense and usually fine to coarse-grained.

Dolomite has other important distinguishing criteria too, analogous to calcite. These include its structural fineness, texture, and type of constituents. The details have already been discussed in Chapter 2.1.1. Like calcium carbonate, dolomite is also classified by its purity. ISO 3262 part 7, defines just three categories from A to C.

In rare cases, magnesium may be partially or fully substituted by other cations to produce distinctive varieties of dolomite like ankerite [$CaFe(CO_3)_2$] or kutnahorite [$CaMn(CO_3)_2$].

Dolomite deposits can be found across many European countries such as France, Germany, Greece, Italy, Norway, Poland, Spain and Sweden. Dolomite is also processed on other continents – America, Asia, Africa – for example in countries like Indonesia, Nigeria and Argentina.

2.2 Silicas

The second most abundant inorganic compound on planet Earth is silicon dioxide, or silica. The English-speaking world considers it a silicate, although German textbooks regard silica as an oxide. The chemical formula for the principal oxide of silicon is SiO_2. Silica occurs in many forms, usually crystalline and occasionally in an amorphous state. Nine separate crystalline forms exist, the most frequently encountered being quartz, cristobalite und tridymite. Siliceous rocks are another form of silica.

Table 2.3: Physical properties and identification numbers of selected silicas used by the filler industry

Mineral	Chemical formula	CAS no.	Density [g/cm^3]	Refractive index	Mohs hardness
Quartz	SiO_2	14808-60-7	2.65	1.54	7
Cristobalite	SiO_2	14464-46-1	2.32	1.49	6.5
Kieselguhr	SiO_2	61790-53-2	1.95	1.46	5.5
Siliceous earth	SiO_2	68855-54-9	2.6	1.55	5.5
Neuburger siliceous earth	SiO_2	1020665-14-8	2.6	1.55	7/2.5*

* Content silica/kaolinite

These contain over 50 percent amorphous and crystalline opals and varieties of quartz, for example siliceous earths, and are industrially significant depending on the extent to which they have been compressed. Diatomite, also known as kieselguhr, is one example of a highly compressed siliceous rock, which generally consists of microquartz.

2.2.1 Quartz

The most important modification of silica is definitely quartz [9–11]. It is found in SiO_2-rich magmatic rocks like granite, and metamorphic rocks like gneiss and quartzite. Other deposits exist as gangue quartz (ore) in sediments like silica sand and flint, as well as in sedimentary rocks (siliceous rocks, sandstones). From the visual aspect, quartz is usually colourless or white, from transparent through cloudy to opaque. Reddish, yellowish, brownish and other colorations indicate trace element impurities, created in part by radioactive bombardment. The concentration of trace elements is less than 100 ppm [12] as a rule. Inclusions of other minerals, like golden needles of rutile, tourmaline, goethite and haematite are especially frequent in varieties of quartz such as rock crystal. Milky quartz, rose quartz, aventurin quartz, chalcedony, agate, onyx, jasper and opal are some more well known varieties. Quartz is categorised in purity classes A and B according to ISO 3262 part 13.

Crystalline quartz occurs in two modifications. α-quartz with its trigonal crystal system is the most prevalent. At temperatures above 573 °C it turns into β-quartz. The process is reversible on cooling. This phase change is in fact used as a "geological thermometer" [13–15]. Figure 2.6 shows other stable zones for silicon dioxide modifications. Coesite and stishovite are two high-pressure modifications that could be produced in theory, but are insignificant as raw materials.

The α-quartz tectosilicate consists of SiO_4 tetrahedrons interlinked by oxygen bridges to produce pseudohexagonal staggered O-rings. Transition to the β-quartz modification simply involves widening the lattice, in other words the O-rings are stretched. Intergrowths are very frequent in alpha quartz, and these lead to twin crystals. Fibrous quartz is a columnar, dense, grained or fibrous aggregate of quartz.

Diatomaceous earth is one of many quartz varieties. It consists largely of SiO_2, but often occurs as a mineral blend of quartz and kaolin. As a result, there are also small quantities of iron oxide, titanium dioxide and calcium/magnesium oxide present. Because of the occasionally high kaolin content, the term diatomaceous earth is used interchangeably for kaolins that contain quartz, for kieselguhr and related sediments. There is a well-known deposit of diatomaceous earth at Neuburg an der Donau in Germany, composed of 60 to 90 % siliceous earth and 40 to 10 % kaolin.

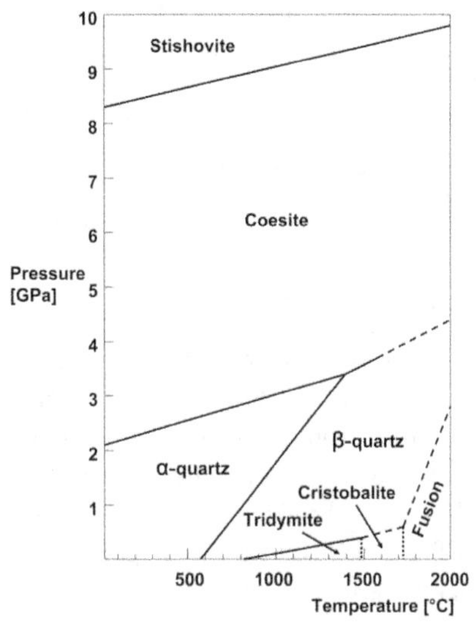

Figure 2.6: Pressure-temperature diagram for modifications of silica

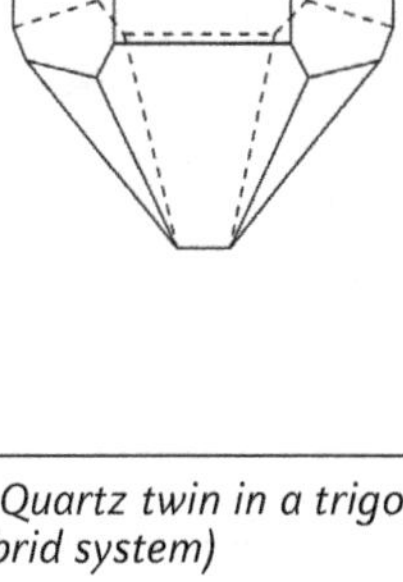

Figure 2.7: Quartz twin in a trigonal crystal system (hybrid system)

2.2.2 Cristobalite

Cristobalite [16] is a mineral that occurs naturally in volcanic rocks and basalts, for example in the Eifel region of the Rhineland Palatinate, and in lunar basalt. In its natural form, it is insignificant as a filler. Cristabolite is turbid with a milky-white colour, and has a tetragonal crystal system.

Cristobalite for use as a filler is made synthetically by thermal modification of quartz.

Figure 2.6 shows the phase change from alpha to beta quartz. By adding further thermal energy, β-quartz turns into β-cristobalite at 1,027 °C. Transformation into α-cristobalite occurs on subsequent cooling to 180 to 270 °C. The resulting SiO_2 is at least 99 % pure as a rule, although this also depends on the purity of the original quartz. Traces of aluminium and iron oxide are possible. ISO 3262 part 14 defines a minimum cristobalite content of 60 %, with 98 %-plus silica. A more precise classification, like that in general use for mineral fillers, does not exist.

2.2.3 Kieselguhr

This amorphous mineral goes by a variety of names. Apart from Kieselguhr [17, 18], diatomaceous earth (latin: terra silicea) and diatomite are also in common use. Kieselguhr was formed from the seabed sediment of tiny algae remains (diatoms) in the Triassic period. These microscopic life forms, less than 300 µm in size, have highly shaped silicic acid frameworks like disks, cylinders and stars, with numerous deepenings, channels and very fine grooves. This rather special framework gives kieselguhr its low density and fine-grained, loose, chalk-like appearance. The framework is often voluminous and porous, which also results in a high specific surface area of 10 to 20 m^2/g. Natural deposits of kieselguhr consist of 70 to 90 % amorphous opal A, 3 to 12 % water, plus organic admixtures. Analysis may also reveal traces of iron, aluminium, calcium, magnesium, manganese, titanium, sodium, potassium, phosphorus and sulphur. The naturally occurring powder is not especially white, and is often light grey in colour. It is also found with a greenish or reddish tint. For this reason, kieselguhr is frequently calcined or flux calcined to give a whiter, purer product. However, thermal processing does not affect its external structure, chemical inertness or porosity, which is due to capillaries and cavities in the silicic acid framework. ISO 3262 part 22 provides information on the requirements for flux calcined kieselguhr. Economically viable deposits are currently being worked in certain states of the USA, for example Lompoc in California. Other mining areas include Queretario in Mexico, the Auvergne region of France, Siebenburgen in Romania, Spain, Denmark, the CIS states, and South Korea.

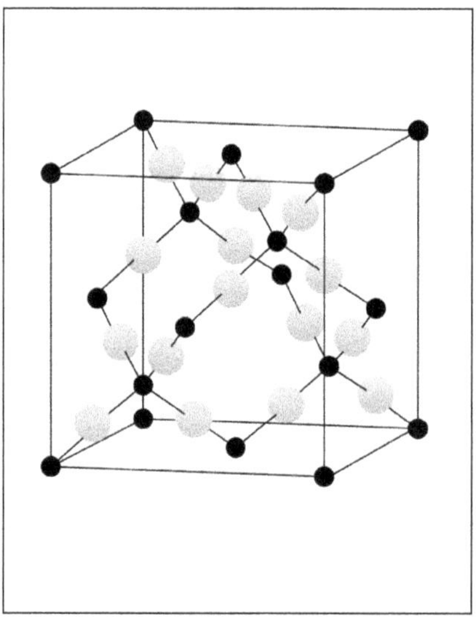

Figure 2.8: Cristobalite crystal system

Table 2.4: Structural classes of silicates, with examples

Silicate class	Structure	Example	Chemical formula
Nesosilicates	Island silicate with one anion $[SiO_4]$	Phenakite Olivine Zircon	$Be_2[SiO_4]$ $(Mg,Fe)_2[SiO_4]$ $Zr[SiO_4]$
Sorosilicates	Group silicate: $[SiO_4]$ tetrahedron with defined groups	Thortveitite Hemimorphite	$Sc_2[Si_2O_7]$ $Zn_4[(OH)_2/Si_2O_7] \cdot H_2O$
Cyclosilicates	Ring silicate: rings of $[SiO_4]$ tetrahedrons	Benitoite Beryll	$BaTi[Si_3O_9]$ $Be_3Al_2[Si_{16}O_{18}]$
Inosilicates	Chain silicate, belt silicate: unlimited single to quintuple chains and belts of anions	Tremolite (amphibole) Sodium waterglass	$Ca_2Mg_5[(OH)_2/(Si_4O_{11})_2]$ $Na_2(SiO_3)\infty$
Phyllosilicates	Sheet silicate, layer silicate: $[SiO_4]$ tetrahedrons chained together in a single plane → layer lattice	Talcum Kaolinite Muscovite	$Mg_3[(OH)_2/Si_4O_{10}]$ $Al_4[(OH)_8/Si_4O_{10}]$ $KAl_2[(OH)_2/AlSi_3O_{10}]$
Tectosilicates	Framework silicate: $[SiO_4]$ tetrahedrons chained in three spatial dimensions → networks	Albite (feldspar)	$Na[AlSi_3O_8]$

2.3 Silicates

Silicates are the most diverse class of minerals around, and hold extraordinary geological and technical significance. Silicates make up over three quarters of the Earth's crust. They come in numerous varieties, and there are various approaches to classifying them: for example according to their degree of dispersion in coarse (minerals), colloidal (clays) and molecular dispersed (in strongly alkaline solutions) silicates. Looking at the silicates' structure is another approach. They are all based on the same, simple scheme of one silicon atom surrounded by four atoms of oxygen; the only differences are in the edge bonds of these SiO_4 units. It follows that there are really just six main kinds of silicate [19].

Silicate crystals that have a belt structure may be fibrous, like hornblende asbestos, or columnar, like actinolite. There is a quite extensive interaction between a given silicate's crystal structure and its physical-chemical properties. Platy structured mica also happens to flake easily. Yet in island-structured silicates, it is just the opposite. Belt structured clays swell as they take in water between their silicon-oxygen layers. Table 2.5 lists silicates of major significance in the filler industry. Kaolinite, muscovite and albite have been chosen for being exemplary of kaolin, mica and feldspar.

Table 2.5: Physical properties and identification numbers of principal silicate fillers

Mineral	Chemical formula	CAS no.	Density [g/cm³]	Refractive index	Mohs hardness
Talcum	$Mg_3[(OH)_2/Si_4O_{10}]$	14807-96-6	2.8	1.57	1
Kaolinite	$Al_4[(OH)_8/Si_4O_{10}]$	1318-74-7	2.6	1.57	1
Muscovite (mica)	$KAl_2[(OH)_2/AlSi_3O_{10}]$	1318-94-1	2.8	1.58	2 to 2.5
Albite (feldspar)	$Na[AlSi_3O_8]$	12244-10-9	2.6	1.54	6 to 6.5

2.3.1 Talcum

Talcum [20–22] (or talc) is a common, naturally occurring magnesium silicate hydrate with the chemical formula $Mg_3[(OH)_2/Si_4O_{10}]$. Talcum was formed several million years ago by hydrothermal metamorphosis of ultra-basic, magnesium-rich rocks like peridotites, especially by low-temperature regional metamorphosis. It is this same magnesium metasomatosis that can turn dolomite or quartzite into talcum. Talcum's composition varies considerably from one deposit to another, because it is always found in combination with other minerals. Here are four typical derivatives that are known precursors to talcum formation:

- Magnesium carbonate, occasionally with silica (→ steatite talc);
- Serpentine (→ soapstone);
- Quartzite, shale or gneiss;
- Sedimentary magnesium rocks.

Talcum occurs most frequently in combination with magnesium aluminium silicate hydrate chlorite, with the chemical formula $(Mg_{6-x-y}Fe_yAl_x)(Si_{4-x}Al_x)O_{10}(OH)_8$. For this reason, ISO 3262 part 10 defines classifications A through D based on the talcum/chlorite content. Carbonate-containing talcum is classified as grade A or B according to ISO 3262 part 11. Percentages of accessory minerals like chlorite, calcite, dolomite, magnesite, siderite, quartz, serpentine, forsterite, magnetite, hematite, pyrrhotine, feldspar, tremolite and anthophyllite may also be specified. Material containing less than 50 % talcum is still deemed a talc-containing mineral, but not actually talcum.

Talcum belongs to the phyllosilicate class of silicates (triple-layer silicates), which means its plate structure is mica-like. It has a crystal lattice formed of $[Si_2O_5]_\infty^2$ double layers, their planes interspaced by $Mg(OH)_2$ units comprised of a central Mg^{2+} ion surrounded by an octahedral arrangement of four oxygen atoms and two hydroxyl groups.

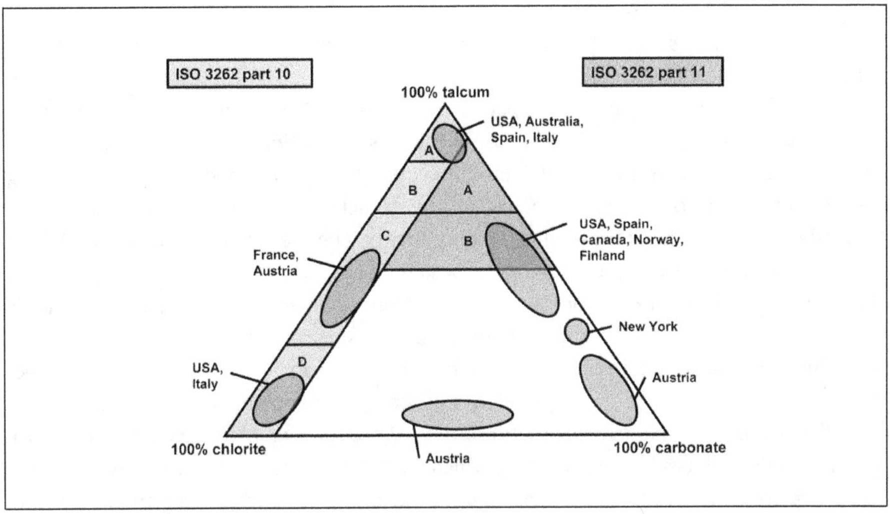

Figure 2.9: Talcum deposits and their mineral composition according to ISO 3262 parts 10 + 11

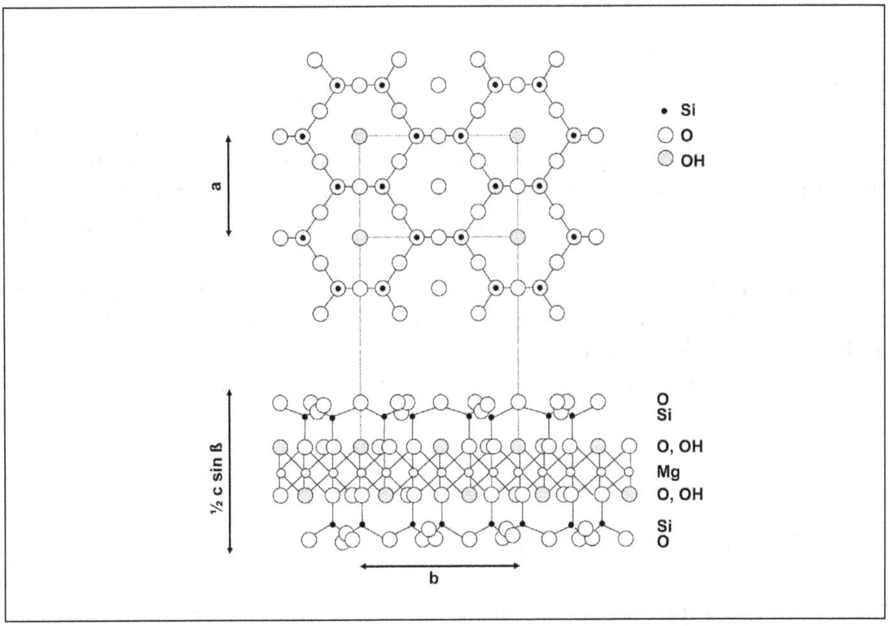

Figure 2.10: Crystal structure of talcum (layer and plane)

A typical talcum might contain 61 % SiO_2, 31 % MgO, 5 % H_2O, 1.4 % Al_2O_3, 1.1 % FeO, along with traces of CaO, CO_2, manganese, titanium, chromium, nickel, sodium and potassium.

Talcum rocks may range from opaque to transparent, and from colourless through white to pale green in appearance. Fe^{2+} or Ni^{2+} ions substituting for Mg^{2+} ions in the crystal lattice impair the brightness of talcum fillers: the mineral may seem white, but milling quickly turns it grey or yellow-brown. Using iron or nickel in the dressing process can significantly reduce the interfering mineral components. There are only weak Van der Waals forces stabilising the talcum double layers, which means they can easily slide against one other and split off (delamination). Moreover, talcum behaves in a strongly hydrophobic manner – although this is only true at the boundary surfaces vertical to the layer plane, not at rupture zones. Talcum occurs in macro- or microcrystalline form, which has a direct effect on the ratio of plate length to plate thickness, also known as aspect ratio. Macrocrystalline talcum has a high aspect ratio, with very pronounced plates. By contrast, microcrystalline talcum has a low aspect ratio, which considerably attenuates its platy structure.

There are extensive talcum deposits in the Austrian province of Steiermark; Pinerolo in Italy; Trimouns in France; Haicheng in China; in Spain, Finland, Norway, Ireland, the Shetland Islands, Canada, Australia, Brazil, as well as the North American states of Texas, California, Montana and New York.

2.3.2 Kaolin

Kaolin [23–25], a silicate mineral, is really a collective term for clay rocks and clay earth. The word kaolin is of Chinese origin, derived from kao ling (the name of a high hill outside the city of Ching-te Chen in Jiangxi Province). Other names for kaolin are: argilla, bolus, China clay and porcelain earth. The minerals kaolinite (often colloquially known as kaolin), halloysite and dickite are regarded as kaolins.

A good way to distinguish kaolins is by the mechanism of their formation. They originate from high-feldspar rocks like granite, rhyolite and arkose. These rocks were hydrothermally transformed, or weathered, under moist, acidic climate conditions (CO_2 dissolved in water) in a process known as kaolinisation. Equation 2.2 shows its chemical scheme.

$$2K[AlSi_3O_8]+7H_2O \rightarrow Al_2O_3 \cdot 2SiO_2 + 4H_2SiO_3 + 2KOH$$

Equation 2.2 : Summary equation exemplifying the kaolinisation process

Residual kaolins exist as contracted, shallow materials, the result of heavy chemical weathering on existing and fossil land areas in the moist, warm tropics and sub-tropics. The original rock structure remains clearly visible. Deposits of this raw material can be found in Germany at Halle, Seilitz near Meißen, Kemmlitz and Hirschau, as well as at Pilsen in the Czech Republic. Hydrothermal kaolins took shape as aisle and tube shaped bodies through hydrothermal transformation of the bedrock [26, 27]. There are well-known deposits in Cornwall, England. On continental landmasses, kaolinitic weathered crusts on the peripheries of lakes, brackwater and rivers produced deposits of weathered-away material known as kaolinitic earths. These are distinguished into varieties of plastic kaolinitic earth like fire clays (sometimes known as refractory clay) and ball clays. Minor impurities present can include mica, illite, smectite, anatase, iron minerals or quartz. There are plastic kaolinitic earth deposits in the Paris basin of France; at Westerwald, Oberpfalz and Lausitz in Germany; and in Cornwall, England. Flint clays belong to the category of non-plastic kaolinitic earths and were formed from lateritic rocks. Such deposits often contain aluminium hydroxide minerals as well. Industrially mined deposits exist in the North American state of Missouri; near Prague in the Czech Republic; as well as in Japan and China. Flint clays were once used for making fireproof goods, which is why to this day they are often called porcelain earths or China clays.

Kaolinite, with the chemical formula $Al_4[(OH)_8/Si_4O_{10}]$, is a triclinic, two-layered clay, and a phyllosilicate. Kaolin is largely comprised of fine-flaked aggregate and dense, mealy or crumbly clumps of kaolinite that may be white, yellowish, brownish or grey in appearance. Kaolinite is structured from alternating layers of $[SiO_4]$ tetrahedrons and $[Al(O,OH)_6]$ octahedrons (hydrargillite layer); see Figure 2.11. The sandwiched layers hold together by hydrogen bonds and dipole interactions.

The composition of natural kaolins varies, depending on the deposit. Hence like talcum, they are classed A through D by kaolinite content. Classification is defined in ISO 3262 part 8.

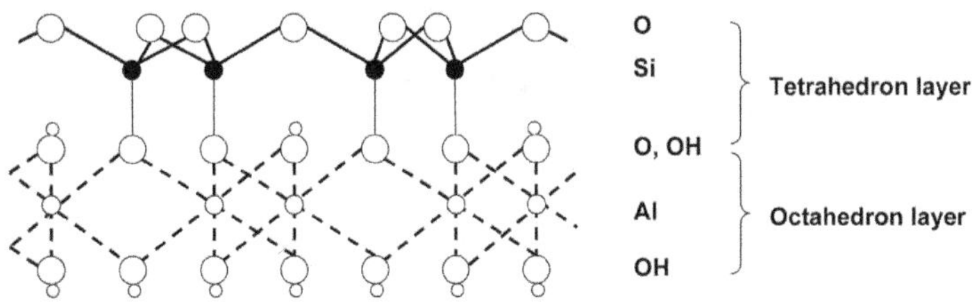

Figure 2.11: Crystal lattice structure of a single kaolinite layer

2.3.3 Mica

Mica is a phyllosilicate, just like talcum and kaolin [28, 29]. It is made up from thin, flexible, springy sheets. The generalised chemical formula of hydroxy- and alkali-containing clay earth silicates is $W(X,Y)_{2-3}[(OH,F)_2/Z_4O_{10}]$, whereby W is usually potassium, sodium or calcium. (X,Y) is normally formed from two ions each of aluminium, magnesium and/or iron or lithium. Z is usually silicon and aluminium. That makes $KAl_2[(OH)_2/AlSi_3O_{10}]$ the chemical formula for muscovite. Together with biotite, muscovite is the most commonly occurring form of mica. Both minerals can be found in magmatic rocks like pegmatite and granite, as well as metamorphic rocks like gneiss and shale slate. There are commercially mined deposits of muscovite in the Canadian province of Ontario, the North American states of New Hampshire and South Dakota, as well as in India, Madagascar, Zimbabwe, Brazil, Norway, and Russia. On account of its colour, which ranges from white to silver-grey but can also be reddish or greenish, muscovite is the sole mica mineral with use as a filler. ISO 3262 part 12 defines the composition of muscovite mica with tolerance bands for oxides of potassium, aluminium and silicon, as well as maxima for iron and magnesium oxides.

Accessory minerals like quartz and chlorite may also occur in varying concentrations along with the muscovite. A composition of approximately one-third each of muscovite, quartz and chlorite represents a special mineral intergrowth known as plastorite. This mineral is a compound of nearly cubic quartz and two layer silicates.

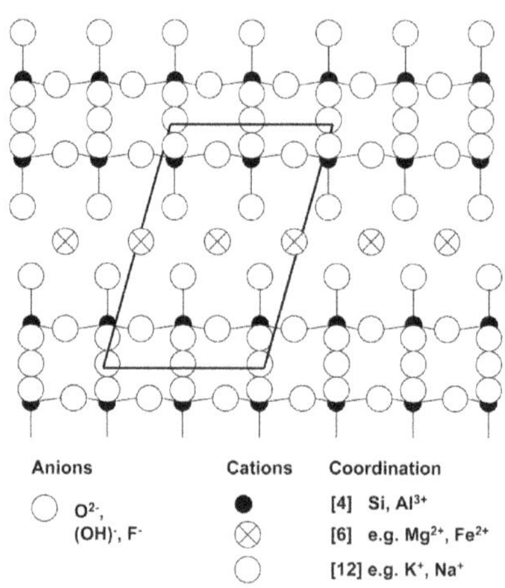

Figure 2.12: Crystal lattice structure of mica (sandwiched layers)

Mica belongs to the triple-layer silicates. These form predominantly monoclinic, usually six-sided (pseudohexagonal) crystals. Their structure consists of two $[(Si,Al)O_4]$ tetrahedral layers with hexagonal symmetry, and embedded between them an octahedral layer of (X,Y) cations. The double-layered mica lattice is similar to that of talcum, with one exception: in talcum, the tetrahedrons are made solely of silicon ions; in mica, every fourth Si^{4+} ion is replaced by an Al^{3+} ion. This produces electrostatic forces between adjacent double layers, which explains why mica is a harder material than talcum. Moisture expansion can cause the mica lattice structure to distend, with alkalis being replaced to some extent by water molecules. The result is hydrous mica. For example, muscovite then forms illite (hydromuscovite).

2.3.4 Feldspar

Feldspars are the most common group of minerals by far [30, 31]. They make up more than 60 % of the Earth's crust, occurring in virtually all magmatic and metamorphic rocks, as well as in many sedimentary rocks. Feldspar is an alumosilicate, composed of a ternary system of three minerals with very similar crystal structures: orthoclase, $K[AlSi_3O_8]$; albite, $Na[AlSi_3O_8]$; and anorthite, $Ca[AlSi_2O_8]$. Minerals whose composition is comprised between albite and anorthite are known as plagioclase feldspars, while those comprised between albite and orthoclase are called alkali feldspars.

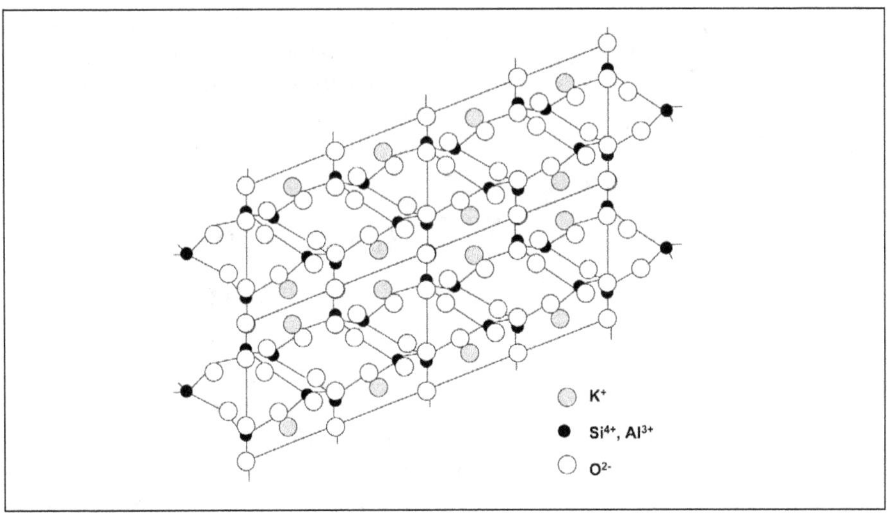

Figure 2.13: Crystal lattice structure of feldspar

Eurofel, the European Association of Feldspar Producers [32], places feldspars in three families, depending on their chemical composition. Feldspar itself is defined as having a silica content in excess of 58 %, alkali oxides (Na_2O+K_2O) above 6 %, and more than 14 % aluminium oxide (Al_2O_3). Material with an aluminium oxide content between 14 % and 8 % is pegmatite. A content of Al_2O_3 lower than this, but above 3 %, is deemed feldspar sand by the Association. Material with an Al_2O_3 content below 3 % does not meet the definition for feldspars.

Feldspars are tectosilicates. Their basic structural building blocks are four-membered rings of [SiO_4] and [AlO_4] tetrahedrons. Those tetrahedrons, chained together via shared oxygen atoms, form three-dimensional networks. Filling the voids in this framework are K^+, Na^+ and Ca^{2+} cations. Allocation of silicon and aluminium in the tetrahedrons is very much the consequence of how the feldspar was formed [33]. Depending on the temperatures involved, there once existed high and low-temperature states that induced growth of either monoclinic or triclinic crystal systems. Formation temperature also governs the miscibility of alkali feldspar with plagioclases, around 900 °C being the optimum. At 600 °C, alkali feldspars exhibit gaps in miscibility. Crystals of feldspar have a prismatic character. They range from the microscopic to the massive in size, with larger specimens often occurring as twin crystals. Feldspars are typically almost opaque, with a grey-white, reddish or yellowish colour. Nepheline syenite is an anomaly among the feldspars. The structure of this tectosilicate is essentially identical to feldspar, but differs in its ratio of aluminium to silicon (1:1), which is considerably higher than that for the feldspars orthoclase and albite. Relative to North America, products based on nepheline syenite are hardly widespread in Europe. There are industrially significant feldspar deposits in Italy, Turkey, the USA, Thailand, Germany, France, Sweden and Norway.

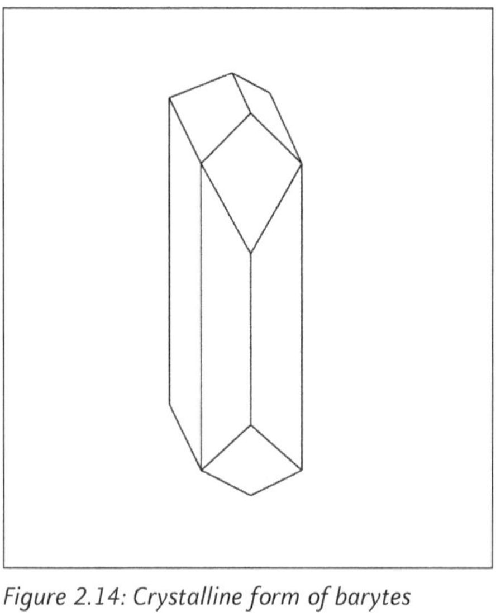

Figure 2.14: Crystalline form of barytes

2.4 Barium sulphate

Natural barium sulphate [34, 35] is known as barytes (also barite and barites) and has a high density, in excess of four grams per cubic centimetre. The conditions for its formation are exclusively low temperatures, as found in mesothermal and epithermal deposits, plus terrestrial to marine sediments. Barium sulphate deposits are very often found on medium and low-temperature hydrothermal veins. The mineral frequently occurs in conjunction with lead, silver and antimony sulphides. Barium sulphate can also occur as replacement deposits in lime and dolomite (as a karst appearance). Chemically deposited barium sulphate can be found at thermal springs, in iron-manganese rich jasper, and in cavities within basalt rocks. The desert, or sand rose, is a rare but well-known form of barytes. Of course, deposits of natural barium sulphate have varying compositions, which is why the resulting filler products are graded by purity into classes A ($BaSO_4$ content >90 %) and B ($BaSO_4$ content >80 %). Comprehensive classification details can be found in ISO 3262 part 2.

Rhombohedral (dipyramidal) barytes crystals are usually large and well shaped. Despite the many different forms afforded by this crystal structure, barytes almost always occurs in a slab shape, or as a finely lamellar twin. Prismatic or columnar crystals are rarer. On the other hand, coarse-flaked and rosette-like aggregate formations are very typical. Barytes can form mixed crystals, for example with coelestin.

The most significant barytes deposits are in China, Russia, Mexico, India, Turkey, the USA, Germany, Morocco, Ireland, Thailand, France, Italy and Brazil. There is a remarkable deposit at Meggen in Germany, where hydrothermal sedimentation has yielded massive barytes, albeit with a grey-black colour.

2.5 References

[1] Geyssant, J., Geological History of Calcium Carbonate, p. 26–29, Plüss-Staufer AG, Oftringen, 1993
[2] Tegethoff, F. W., Calcium Carbonate – From the Cretaceous Period into the 21st Century, p. 16 ff, Birkhäuser Verlag, Basel, 2001
[3] CCA-Europe – The European Calcium Carbonate Association, Internet www.ima-eu.org/en/ccaindex.html, 2004
[4] Kittel, H., "Lehrbuch der Lacke und Beschichtungen – Pigmente, Füllstoffe und Farbmetrik", vol. 5, p. 431–439, S. Hirzel Verlag, Stuttgart, 2003
[5] Wypych, G., "Handbook of Fillers", p. 48–57, ChemTec Publishing, Toronto, 1999
[6] Skinner, A. J., Lafeminai, J. P., Jansen, H. J. F., "Structure of bonding of calcite: A theoretical study", American Mineralogist, vol. 79, p. 205 ff, Mineralogical Society of America, Washington, D.C., 3/1994
[7] Billemann, C., Gillet, P, "High-pressure and high-temperature behaviour of calcite, aragonite and dolomite: A Raman spectroscopic study", European Journal of Mineralogy, vol. 4, p. 389–393, Schweizerbart, Stuttgart, 1992
[8] Wypych, G., "Handbook of Fillers", p. 84, ChemTec Publishing, Toronto, 1999
[9] Eurosil – The European Association of the Silica Sand Producers, Internet www.ima-eu.org/en/eurosil.html, 2004
[10] Kittel, H., "Lehrbuch der Lacke und Beschichtungen – Pigmente, Füllstoffe und Farbmetrik", vol. 5, p. 361–363, S. Hirzel Verlag, Stuttgart, 2003
[11] Wypych, G., "Handbook of Fillers", p. 142-145, ChemTec Publishing, Toronto, 1999
[12] Heaney, P. J., Veblen, D. R., "Observations of the α-β-phase transition in quartz: A review of imaging and diffraction studies and some new results", American Mineralogist, vol. 76, p. 1018–1032, Mineralogical Society of America, Washington, D.C., 6/1991
[13] Carpenter, M. A., "Equilibrium thermodynamics of Al/Si ordering in anorthite", Physics and Chemistry of Minerals, vol. 20, p. 1–24, 1993
[14] Hammonds, K. D. et al, "Rigid-unit phonon modes and structural phase transitions in framework silicates", American Mineralogist, vol. 81, p. 1057–1079, Mineralogical Society of America, Washington, D.C., 9/1996
[15] Xu, H., Haeney, P. J., "Memory effects of domain structures during displacive phase transitions: A high-temperature TEM study of quartz and anorthite", American Mineralogist, vol. 82, p. 99–108, Mineralogical Society of America, Washington, D.C., 1/1997
[16] Wypych, G., "Handbook of Fillers", p. 78–79, ChemTec Publishing, Toronto, 1999
[17] Kittel, H., "Lehrbuch der Lacke und Beschichtungen – Pigmente, Füllstoffe und Farbmetrik", vol. 5, p. 363–364, S. Hirzel Verlag, Stuttgart, 2003
[18] Wypych, G., "Handbook of Fillers", p. 80–83, ChemTec Publishing, Toronto, 1999
[19] Liebau, Structural Chemistry of Silicates, Springer Verlag, Berlin 1985
[20] Eurotalc – The Scientific Association of Talc Producers, www.ima-eu.org/en/talcindex, 2004
[21] Kittel, H., "Lehrbuch der Lacke und Beschichtungen – Pigmente, Füllstoffe und Farbmetrik", vol. 5, p. 371–377, S. Hirzel Verlag, Stuttgart, 2003
[22] Wypych, G., "Handbook of Fillers", p. 150–153, ChemTec Publishing, Toronto, 1999
[23] EKA – The European Kaolin Association, Internet www.ima-eu.org/en/kaolin.html, 2004
[24] Kittel, H., "Lehrbuch der Lacke und Beschichtungen – Pigmente, Füllstoffe

und Farbmetrik", vol. 5, p. 377–383, S. Hirzel Verlag, Stuttgart, 2003
[25] Wypych, G., "Handbook of Fillers", p. 99–103, ChemTec Publishing, Toronto, 1999
[26] Satokawa, S. et al, "Effects of acidity on the hydrothermal synthesis of kaolinite from silica-gel and gibbsite", Clays, vol. 44, p. 417–423, The Clay Minerals Society, Aberdeen, 1996
[27] Huertas, F. J., Huertas, F., Linares, J., "Hydrothermal synthesis of kaolinite: Method and characterization of synthetic materials", Applied Clay Science, Jg. 7, S. 345–356, Elsevier, London, 1993
[28] Kittel, H., "Lehrbuch der Lacke und Beschichtungen – Pigmente, Füllstoffe und Farbmetrik", vol. 5, p. 384–386, S. Hirzel Verlag, Stuttgart, 2003
[29] Wypych, G., "Handbook of Fillers", p. 112–115, ChemTec Publishing, Toronto, 1999
[30] Kittel, H., "Lehrbuch der Lacke und Beschichtungen – Pigmente, Füllstoffe und Farbmetrik", vol. 5, p. 386–389, S. Hirzel Verlag, Stuttgart, 2003
[31] Wypych, G., "Handbook of Fillers", p. 86, ChemTec Publishing, Toronto, 1999
[32] Eurofel – The European Association of Feldspar Producers, www.ima-eu.org/en/felindex.html, 2004
[33] Zhang, M., et al, "Phonon-spectroscopy on alkali-feldspars: Phase transitions and solid solutions", American Mineralogist, vol. 81, p. 92–104, Mineralogical Society of America, Washington, D.C., 1996
[34] Kittel, H., "Lehrbuch der Lacke und Beschichtungen – Pigmente, Füllstoffe und Farbmetrik", vol. 5, p. 460–466, S. Hirzel Verlag, Stuttgart, 2003
[35] Wypych, G., "Handbook of Fillers", p. 36–40, ChemTec Publishing, Toronto, 1999

3 Production of fillers

Fillers are produced by two basic methods. By far the largest quantity of industrial fillers is made by physical/mechanical processing of natural minerals. In very simple terms, this means breaking up rocks and boulders into ever-finer particles. On the other hand, there are synthetic fillers, often with highly functionalised properties. Their production is more complex and specific, with most processes being based around chemical precipitation reactions. The starting materials for synthetic and natural fillers are to some extent identical.

3.1 Production of natural fillers

Turning natural materials into fillers is a multi-step industrial process. But ahead of production, a sizable deposit needs to be found and associated mining rights secured. With the deposit tapped and investments made in plant facilities, actual production can get underway. Regardless of the actual mineral and its chemical composition – calcium carbonate, barium sulphate, kaolin, or talcum – they all pass through basically similar production stages. Some natural minerals undergo additional processing, frequently purification, to enhance their properties. A secondary treatment process, like surface coating, sometimes comes at the end of the production chain.

3.1.1 Prospecting

Preceding the production of natural fillers is an exacting and quite time-consuming phase of prospecting. Having identified a potential mining site, an initial geological survey is conducted to assess the extent and quality of deposits. Meanwhile, enquiries have to be made into government terms and conditions, and the regional infrastructure, all of which will have a significant bearing on any subsequent mining activity. If findings are positive throughout, further geological exploration becomes necessary. A detailed survey will be made of the site, by drilling a dense grid of sample bores. By the time this major undertaking is complete, a lot will be known about the site, including [1]:

- Size of deposit
- Quality and variability
- Thickness and composition of the top layer to be removed
- Geographical extent and dip of the deposit
- Formations like folds and buckles
- Presence and concentrations of undesired foreign minerals

Now is the time to evaluate investment expenditure in mining and production plant, infrastructure, and future logistics in the light of the site's proximity to the market. If the deposit still appears viable for mining, a spatial model is made to reflect the criteria listed above. The model additionally serves to document the way in which mining activity will proceed, and its full impact on the natural surroundings. The prospecting phase is complete when all conditions are met and the necessary official authorisations granted. The process may well take several years, but it is essential before investing resources and, finally, starting filler production.

Figure 3.1: Marble quarry worked in decks source: Omya AG

3.1.2 Mining

With prospecting brought to a successful close, it is time to prepare for mining. The specifics of the preparatory phase depend on the raw material: mining techniques require adapting to the mineral concerned, and to the vagaries of a given deposit. There is an overarching distinction here between opencast and underground mining. As its name suggests, opencast mining takes place "in the open", in other words the mineral is extracted from an exposed pit in the ground. The first step is to

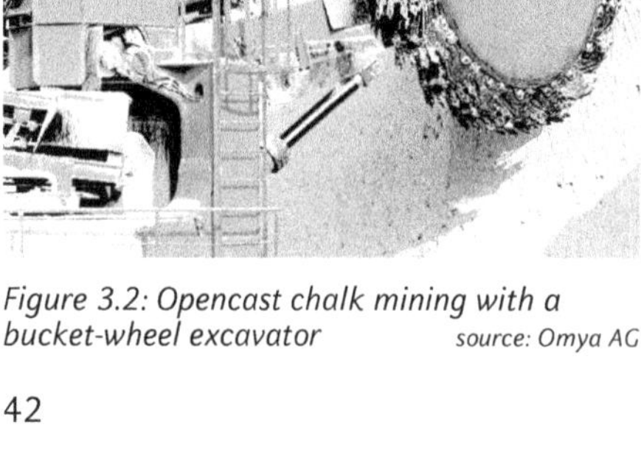

Figure 3.2: Opencast chalk mining with a bucket-wheel excavator source: Omya AG

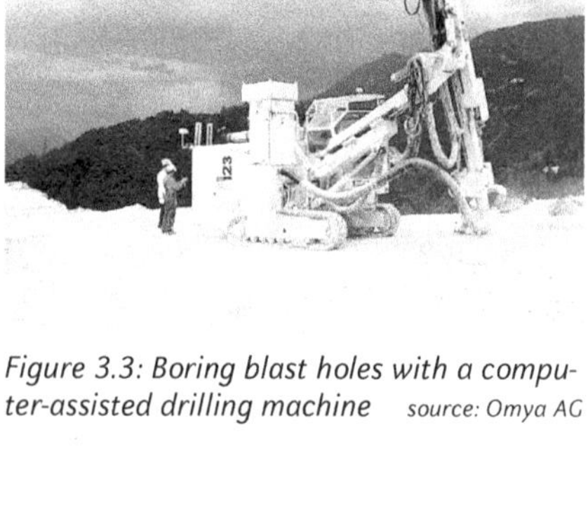

Figure 3.3: Boring blast holes with a computer-assisted drilling machine source: Omya AG

skim off and set aside the top layer, usually consisting of earth and clay. This material will be needed for later re-cultivation, when the mine closes down.

Mining is a purely mechanical process, which is adapted to the specific mineral and deposit. Quarries are worked in multiple decks, to assure their continuous operation. Minerals such as chalk, layer silicates like talcum, kaolin, mica and feldspar, and silica in the form of quartz sand and diatomaceous earths will typically be extracted by machinery like bucket-wheel and shovel excavators. Explosives may also be deployed to dislodge highly compacted rock.

Figure 3.4 to 3.7: Blasting sequence in a marble quarry *source: Omya AG*

Explosives are likewise used for quarrying calcite (limestone and marble), as well as dolomite and barytes. The system of drilling and blasting is calibrated to the rock itself and the type of explosive used. A series of slanting holes with defined bore and spacing is drilled from the upper to the lower quarry deck. Blasting then leaves a clean-cut free face, so further mining activity can proceed without a break. Getting the blast right also produces loose material of a convenient size for loading, transportation and further processing, with no need for breaking into smaller pieces at the quarry face.

Loaders, shovel dozers or excavators place the loose blasted material onto conveyor belts for transportation over longer distances, or heavy trucks for short runs. The raw material is given an initial visual inspection at this stage, with inferior-grade boulders picked out and taken to a waste dump. Satisfactory boulders are transported to primary crushers for the first controlled breaking-up stage.

Figure 3.8: Gargantuan shovel dozers load blasted rock onto 80-ton heavy trucks

source: Omya AG

Figure 3.9: Drilling blast holes underground

source: Omya AG

Underground mining becomes viable where very rich deposits lie beneath an excessively thick top layer. In contrast to opencast mining, underground mining only removes material directly from the deposit. So, substandard material is selectively left in place underground, instead of being taken out to a waste dump. This diminishes the higher cost of underground mining compared to the opencast method. Various means of extraction may be used, depending on the mineral. Mining in lightly compressed rocks requires the use of props and supports; denser rocks can be mined using a chamber process. Indeed, mining hard marble deposits in this way can result in cathedral-like underground vaults. Extraction techniques below ground are similar in principle to the opencast methods described previously, although the machinery used tends to be more compact. Moreover, a greater number of safety installations become necessary to protect personnel and equipment.

3.1.3 Processing

Processing of crude rock begins with pre-crushing in "primary crushers". Boulders inside the crusher, some measuring up to 150 cm, are reduced to particle sizes between 300 and 100 mm by pressure (jaw and cone crushers) or impact (impact crusher).

The next stage is often a coarse sieving to remove soil and similar detritus. Sieving also removes finer mineral fragments, which are fed into a later stage of production. Large fragments are given a secondary crushing by cone, impact, roller or hammer crushers. Depending on type and output fineness, crushers like these can process around 2,000 tons of rock per hour.

Material for producing particle sizes of 500 µm and larger is not crushed or ground any further, but instead fed into tumbler, jigger or vibrating screens or sieving towers. Here, granulate containing a distribution of particle sizes is fractionated according to the respective screen coarseness.

The crushed rock is sieved again, and then segregated for further processing by dry or wet grinding. As before, the fine component is separated and fed in to a later stage of the grinding process. The rock may also be sorted by optical means. Nowadays, this is usually done automatically with the help of digital cameras that identify rocks moving past on a conveyor and compare them against a standard, all in a split second. The modern way of removing a piece that does not meet selected requirements is to blow it off the conveyor with a precise jet of compressed air. The number of crushing stages depends on the crude rock, and the size of the original fragments. Before grinding, the rock is given a thorough washing in drum or vibratory washers. A "dry cleaning" process can also be used in situations where wet washing is not an option. Some kinds of rock are cleaned by simply slurrying them: impurities are separated, and the residual suspension filtered and dried for further processing.

Figure 3.10: Material entering a jaw crusher source: Omya AG

A further washing may be needed subsequent to pre-grinding, depending on the mineral and its source deposit. In the filler production process by dry grinding, a magnetic separator may be used to extract ferrous impurities that would otherwise impair whiteness. Magnetic separators with a field strength of 5tesla are now available for upscale industrial deployment. In the wet grinding filler production process, the second washing is done by flotation, which extracts mineral impurities like graphite. Adding organic reagents to the slurry separates the lighter from the heavier suspended particles, with lighter products being propelled to the surface in a three-phase

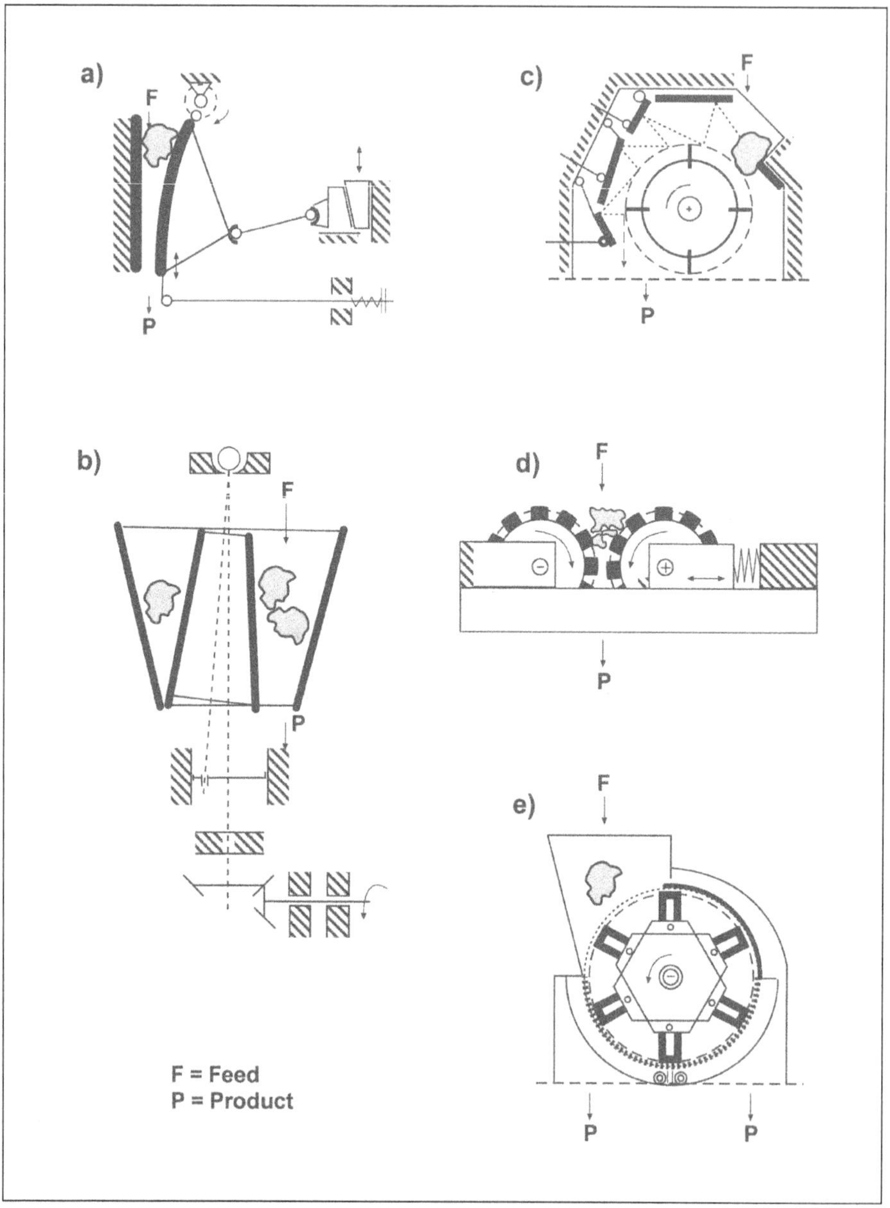

Figure 3.11: Schematic outlines of crusher operation: a) Jaw crusher, b) Cone crusher, c) Impact crusher, d) Roller crusher, e) Hammer crusher source: Omya AG

mixture of water, air and solid matter. This forms foam that is skimmed off the tops of the flotation cells.

Flotation may be followed by dewatering, to revert to a solid material ahead of the final – micronisation – grinding stage that produces the actual filler. Each one of the cleaning and sorting stages described above contributes to enhanced purity, which also increases brightness.

There are two ways of grinding crushed and cleaned minerals. The most widespread method produces dry fillers by the direct route. Dry grinding feed is placed in a drum mill, with steel ball bearings as the grinding medium. This type of grinder, often referred to as a ball mill, is a universal, robust, low-maintenance item of plant for producing micro-fine particles and specific particle size distributions (PSD). Organic grinding agents are added to boost the efficiency of the grinding process – hence reducing its cost – and to prevent filler agglomeration. Ball mill geometry, the grinding medium, the load quantity, and rotational speed are just some of the factors influencing the end result. A ball mill operates at the optimum when the grinding medium is cataracting – that is to say, centripetal force and friction lift a continuous mêlée of grinding medium almost to the apex of a parabolic trajectory, from where it comes crashing back down again. It is the resulting impact and friction that breaks up the grinding feed.

Figure 3.12: Rock washing in a vibratory washer source: Omya AG

Air classifiers, in line with ball mills, are used to separate coarse and fine material

Figure 3.13: Separation of the mineral phase from ores by flotation source: Omya AG

Figure 3.14: Horizontal ball mill for micronising dry fillers source: Omya AG

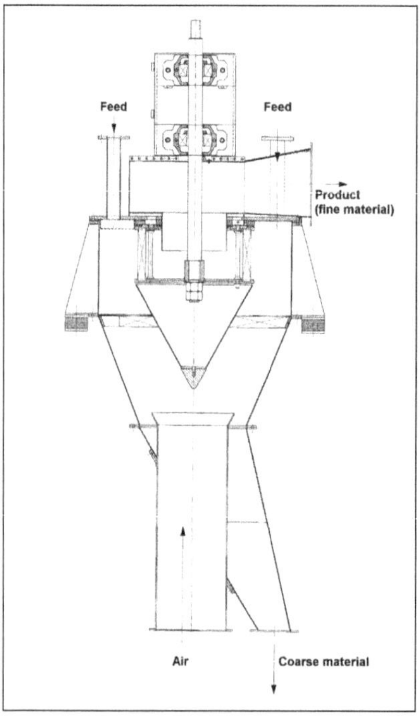

Figure 3.15: Schematic section through an air classifier *source: Omya AG*

from the ball mill. The coarse fraction is returned for re-grinding, thus closing the ball mill circuit. An air classifier works by the stream separation method. Here, ground material comprised of different-sized particles is dropped into a controlled upwards air current. Particles falling at less than the speed of the rising air are swept up with it; those falling faster continue their descent. When the respective speeds coincide, particles hover in suspension. For particles of a given density, the speed of descent depends solely on their diameter. Hence, coarse particles sink while fine particles rise in the updraught. Classification grows more difficult with increasing product fineness, and the energy requirement also rises considerably. Classifiers can also be operated in tandem to yield products with various PSDs. In the finest dry-ground products, 90 % of the constituent particles have a diameter under two micrometres (D90% = 2 µm). For PSDs finer than this, wet grinding finds application on economic grounds: with high-

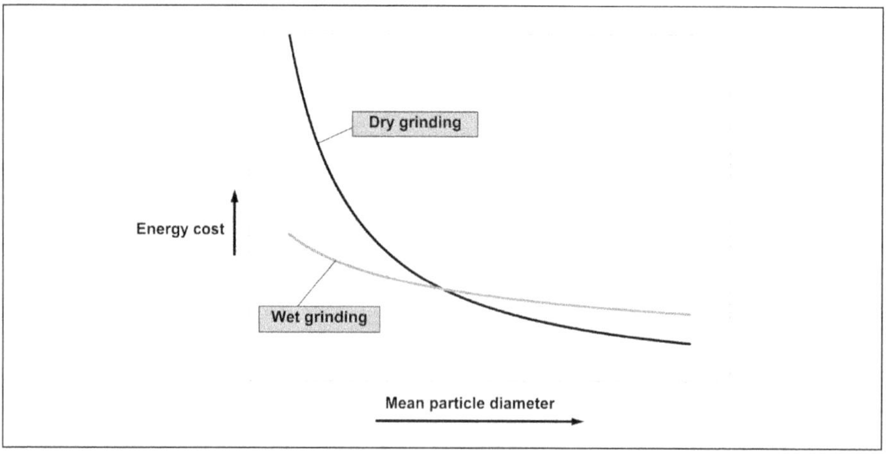

Figure 3.16: Comparative energy costs of dry and wet grinding

fineness to ultra-fine fillers, the energy requirement for wet grinding and subsequent drying is nonetheless considerably below that for dry grinding. However, with median particle diameters in excess of 1.5 µm, dry grinding becomes more economical than wet grinding.

In wet grinding, the grinding feed is put into aqueous suspension with a defined solids content of no more than 75 %, and blended with highly-active dispersants. Grinding is effected in wet-operated agitated ball mills, which are capable of delivering energy sufficient to break up mineral particles to a micrometre or less. The energy requirement, depending on the desired fineness and mode of operation, works out as high as 200 kilowatt-hours per ton of filler. Subsequent drying is also a very energy-intensive process. To reduce costs, the solids content is elevated ahead of actual drying. This preliminary, mechanical dewatering may be done with filters, centrifuges, or a hydrocyclone.

Figure 3.18 shows the overall scheme of the wet and dry grinding steps just described, dry to the left, wet to the right. The paths for these two grinding methods diverge immediately after pre-crushing, where the rock is graded and stored separately.

Thermal processing during production can result in natural fillers with significantly different properties. Classic examples are the thermal processing of kaolin and quartz. When wet-ground (higher purity level) kaolin is exposed to temperatures between 760 °C and 1100 °C in a calcining kiln, it is dehydrated as the water of crystallisation dissociates from crystal lattices, and loses 14 % of its mass in the process. Calcining also oxidises impurities that could not be separated earlier by flotation. Silicon-aluminium spinelles and even mullite may form in the amorphous phase at temperatures above 950 °C. This dehydrated, calcined kaolin is first cooled, and then re-ground to recover the primary particles from aggregates formed by calcining. Compared to the original wet-ground kaolin, this process results in extensively modified properties, like increased oil absorption and brightness.

Figure 3.17: Calcium carbonate in aqueous suspension is ultra-fine ground in agitated ball mills
source: Omya AG

In addition to the classic calcination process, there is flash calcination. Ground kaolin is once again the starting material, but here the product is exposed to a hot gas flame for a few seconds. Water of crystallisation spontaneously dissociates, while the platy structure of the kaolin particles expands. The end result is amorphous particles, enclosing nanoscale internal air cavities. Manufacturing industry puts a premium on this unique structure, the particle shape, and optical properties of "thermo-optic" kaolins as opposed to regular calcined kaolins.

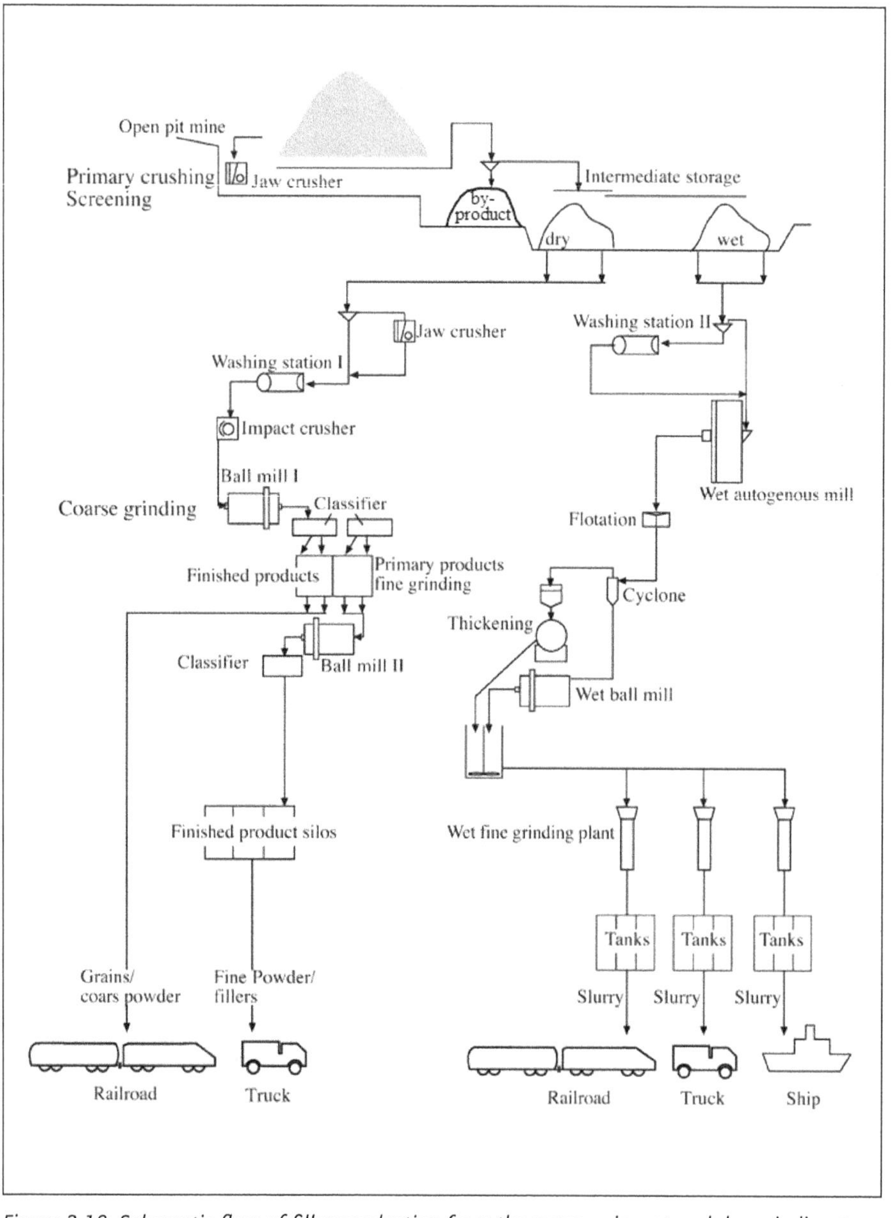

Figure 3.18: Schematic flow of filler production from the quarry, via wet and dry grinding, to final loading
source: Omya AG

Thermal treatment of extremely pure quartz with a silica content of 99 % or more produces cristobalite at temperatures around 1500 °C; see also Figure 2.6. Cristobalite fused from quartz is then quenched and either sieved, or ground and classified.

3.2 Synthetic fillers

Synthetic fillers and functional fillers are frequently associated, yet their manufacturing processes can differ greatly. As well as the processes for respective filler materials, a chemically identical filler can often be produced by more than one process. Fillers are produced synthetically for various technical reasons. For example, the synthetic route can be a means to enhance optical properties – although that depends not only on the process itself, but also on the quality of the starting material. Further to such optical improvements, synthesis can also produce particles of a fineness that would be economically and technically infeasible on an industrial scale by, say, grinding.

3.2.1 Precipitated calcium carbonate

The sole point of similarity between synthetic calcium carbonate and its natural counterpart is in the chemical formula. Its properties are characteristically distinct; hence the existence of a separate standard to define precipitated calcium carbonate. ISO 3262 part 6 describes precipitated calcium carbonate (PCC) as synthetic calcium carbonate comprised of trigonal (calcite) or rhombic bipyramidal (aragonite) crystals. Precipitated calcium carbonate can be produced by various methods [2–6], the most frequent being the direct three-stage process with precipitation by carbon dioxide. Here, very pure limestone is first transformed into calcium oxide, also known as quicklime or simply lime, by heating above 900 °C. The reactive by-product is carbon dioxide, which can be recycled at the third stage of the process. The quicklime is slaked with water to make milk of lime, or chemically speaking calcium hydroxide. This is fed into a reactor as a dilute suspension, and the carbon dioxide produced earlier is passed in until all of the calcium hydroxide has turned into calcium carbonate. The energy balance ΔH for producing quicklime is -3130 kJ/kg CaO; the total energy balance for slaking and precipitation is similarly ΔH = +3130 kJ/kg CaO. The reaction time is monitored and controlled by tracking the change in pH value. Compared to the other processes described below, the direct process is the most efficient and controllable in terms of producing specific properties.

$$CaCO_3 \xrightarrow{>900\ °C} CaO + CO_2 \uparrow$$

Equation 3.1: Step 1 of the direct PCC process

$$CaO + H_2O \longrightarrow Ca(OH_2)$$

Equation 3.2: Step 2 of the direct PCC process

$$Ca(OH_2) + CO_2 \longrightarrow CaCO_3 + H_2O \uparrow$$

Equation 3.3: Step 3 of the direct PCC process

PCC's crystalline form and particle distribution can be altered by adjusting reaction conditions like pressure, temperature, time, and chemical additives. Precipitation at higher temperatures as a rule produces greater amounts of aragonite, and finer particles. Temperatures below 30 °C give an excess of calcite crystals. Agglomerates cannot always be avoided in PCC production, so they are either broken up by subsequent grinding, or separated by a classification process.

Other well-known processes include the Solvay process, the lime soda process, and the reaction of calcium nitrate with ammonium hydroxide in fertiliser production. The

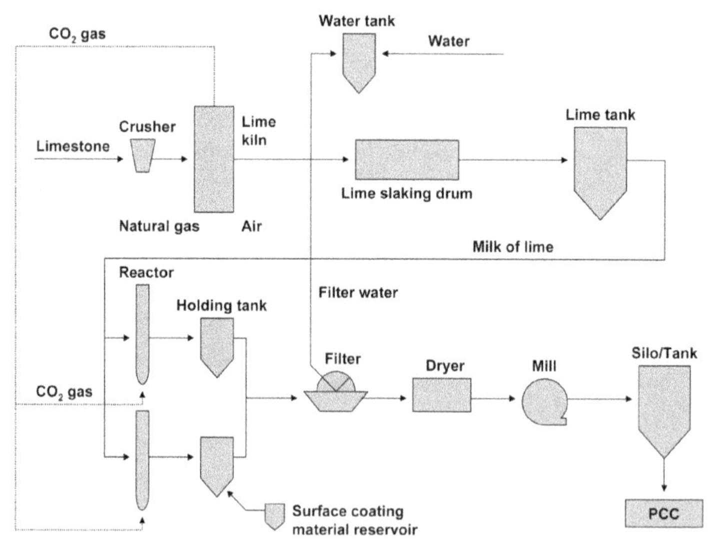

Figure 3.19: Schematic outline for PCC production by the direct process

Solvay process involves, in part, passing carbon dioxide into an ammoniacal solution of calcium chloride; the conversion products are calcium carbonate and ammonium chloride.

$$CaCl_2 + 2NH_4OH + CO_2 \longrightarrow CaCO_3 + 2NH_4Cl + H_2O$$

Equation 3.4: Part of the Solvay process

The lime soda process was originally used to manufacture sodium hydroxide solution, but has been superseded by electrolysis of sodium chloride. Calcium carbonate is obtained as a by-product. As with the direct process, milk of lime is required for the conversion.

$$Ca(OH)_2 + Na_2CO_3 \longrightarrow CaCO_3 + 2NaOH$$

Equation 3.5: Lime soda process

Fertiliser production using calcium phosphate and nitric acid as feedstock yields calcium hydrogen phosphate and calcium nitrate. Passing carbon dioxide through an ammoniacal solution of calcium nitrate makes it react to form calcium carbonate and ammonium nitrate.

$$Ca(NO_3)_2 + NH_4OH + CO_2 \longrightarrow CaCO_3 + 2NH_4NO_3 + H_2O$$

Equation 3.6: Conversion of calcium nitrate to PCC

The reactions illustrated in Equations 3.5 and 3.6 all require lengthy purification processes to ensure consistent product quality.

3.2.2 Precipitated barium sulphate

The synthetic form of barytes is known either as permanent white, or by its French name, "blanc fixe". ISO 3262 part 3 defines blanc fixe as synthetic barium sulphate, made by precipitation. The standard goes on to define a minimum barium sulphate content of 95%, along with maximum permissible concentrations of impurities.

Sparingly soluble barium sulphate is formed in various chemical reactions. What they all have in common is the presence of barium and sulphate ions at the outset of the precipitation reaction. Barium sulphide, barium carbonate, barium chloride, or barium hydroxide may be used as the source of barium ions. The sulphate ion donor is frequently Glauber's salt ($Na_2SO_4 \cdot 10\ H_2O$) or sulphuric acid, or occasionally hydrogen peroxide. The barium sulphate precipitation reaction is a standard in chemical analysis, where it provides a qualitative and quantitative test for barium as well as for sulphate ions.

$$Ba^{2+} + SO_4^{\ 2-} \longrightarrow BaSO_4$$

Equation 3.7: General equation for the barium sulphate precipitation reaction

Pure starting products are vitally important to the quality of the precipitate; impurities can adversely affect both colour and approval criteria for certain applications. Accordingly, only purified products are used in the production of blanc fixe. Product characteristics like particle size and shape are controllable via process parameters: a highly concentrated precipitation solution and higher temperatures will produce very fine particles, while coarser particles are obtained by reducing concentration and temperature. Other influential factors in precipitation include agitation speed, geometry of the precipitation reactor, flow rate, pH, maturing time, presence of foreign ions, etc. Yet precipitation is just a part of the blanc fixe production process. In the following stage, barium sulphate precipitated in solution is purified multiple times by separation (filtration or centrifuge) and by washing

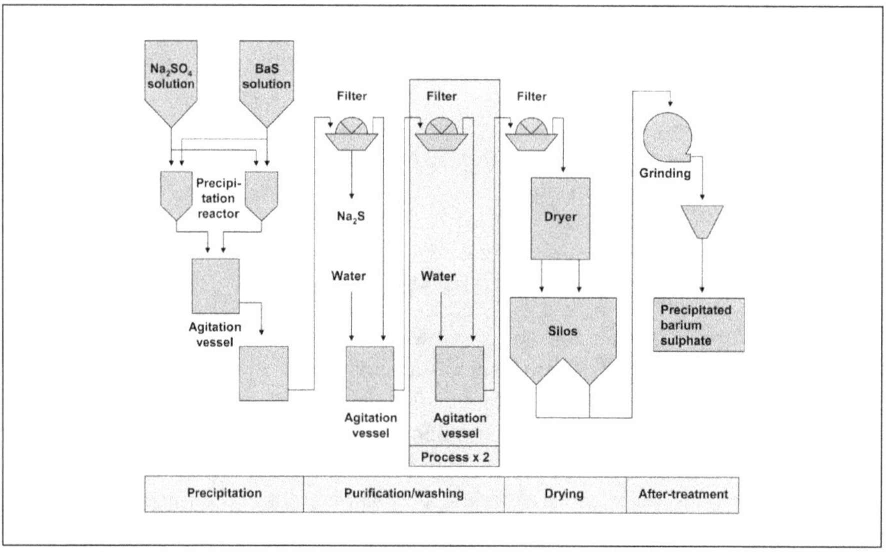

Figure 3.20: Outline for blanc fixe production based on barium sulphide and sodium sulphate

out undesirable components like sodium sulphide, which can give odour problems. In the third stage, blanc fixe for universal use in paints and coatings is then dried and ground to break up agglomerates.

3.2.3 Modified calcium carbonate

Natural calcium carbonate can be further processed as seen in Chapter 3.2.1. The latest development in the field of calcium carbonate is the modification by mechanical and chemical means. Especially, the acid solubility of calcium carbonate is beneficial for the chemical modification. The calcium carbonate is treated with quite strong providers of hydronium (proton) and gaseous carbon dioxide. The present pressure and temperature in the process reactor are further variables for the design of the particle structures. When using such a re-crystallisation process, new calcium carbonate structures can be obtained as a shell on a core, which is still of calcite nature. This process leads to surface modified particles with different functionality compared to GCC.

The modified calcium carbonate (MCC) can offer very high specific surface areas exceeding 100 m²/g characterised by the BET method. Despite the high specific surface area, the particle sizes of MCC stay in the micrometer range. However, one dimension of the particle structure (plate) in the shell can be designed below 100 nm, therefore offer-

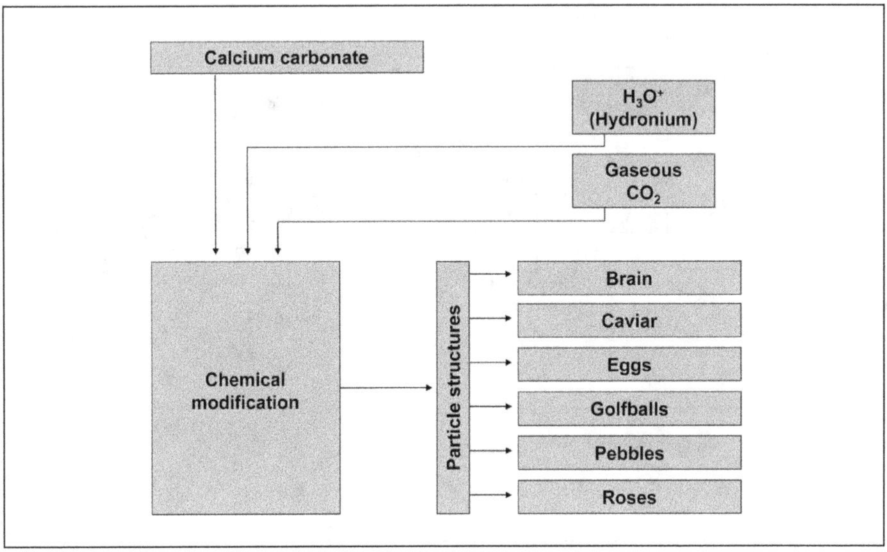

Figure 3.21: Production process for modified calcium carbonate

ing save nano functionality without being a nano particle. The modification process parameters can be adjusted to the requested function and appearance of the particles. This allows the production of amorphous particle structures such as brains, caviar, eggs, golf-balls, pebbles, roses and others. Depending on the particle structure, the specific area is subject to change, see Figure 3.21.

3.2.4 Synthetic silicic acids

Compounds with the general formula $(SiO_2)_m \cdot nH_2O$ are collectively known as silicic acids. The starting point for all synthetic silicic acids is sand that has been processed through various intermediate stages; see Figure 3.22. The reaction of a silicon halide like $SiCl_2$ with water produces mainly orthosilicic acid.

$$SiCl_4 + 4H_2O \longrightarrow Si(OH)_4 + 4HCl$$

Equation 3.8: Formation of orthosilicic acid

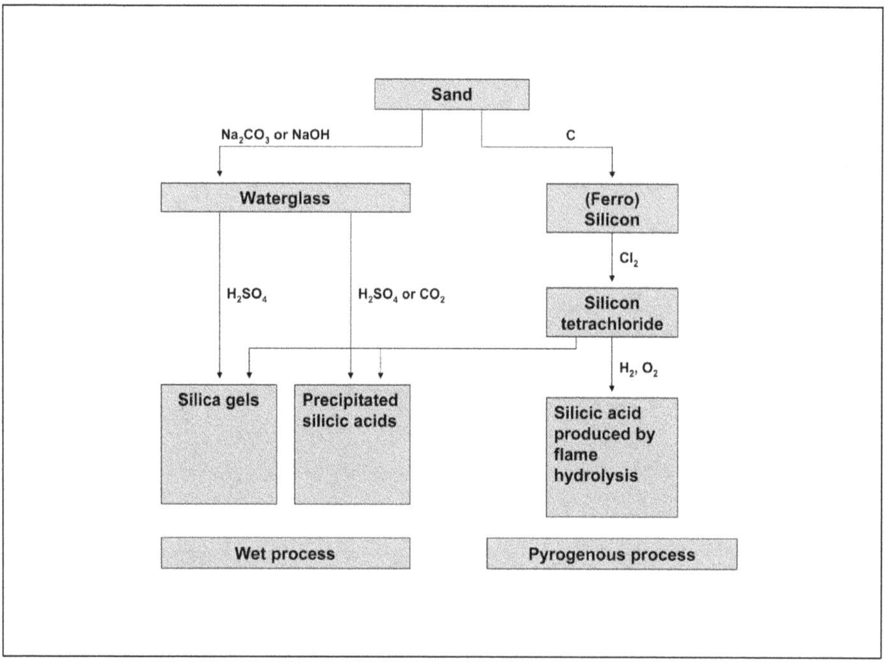

Figure 3.22: Production process for silicic acid

Orthosilicic acid belongs to the group of monosilicic acids. It is a weak acid; heavily diluted and at a pH between 2 and 3 it keeps stable for only a few days. If the pH deviates, intermolecular dehydration occurs. The immediate condensation product is pyrolitic silicic acid (disilicic acid).

$$(HO)_3Si-OH+OH-Si(OH)_3 \xrightarrow[-H_2O]{} (HO)_3Si-O-Si(OH)_3$$

Equation 3.9: Formation of pyrolitic silicic acid as a condensation product of orthosilicic acid

Further condensation can occur in a similar way, resulting in cyclic, cage shaped or almost ball shaped polysilicic acids. Condensation leads to an increase in molecular weight by chain propagation, which can either simply extend the chain, or form rings and branches. The result is an irregular (amorphous) silicic acid structure [7]. The theoretical end product would be a polymeric silica, or silicic acid anhydride. One form of silicic acid is a colloidal solution of polysilicic acids (silica sol), with particle sizes of just 5 to 150 nm. Silica sol is unstable and apt to condense further; aggregation turns it into silica gel.

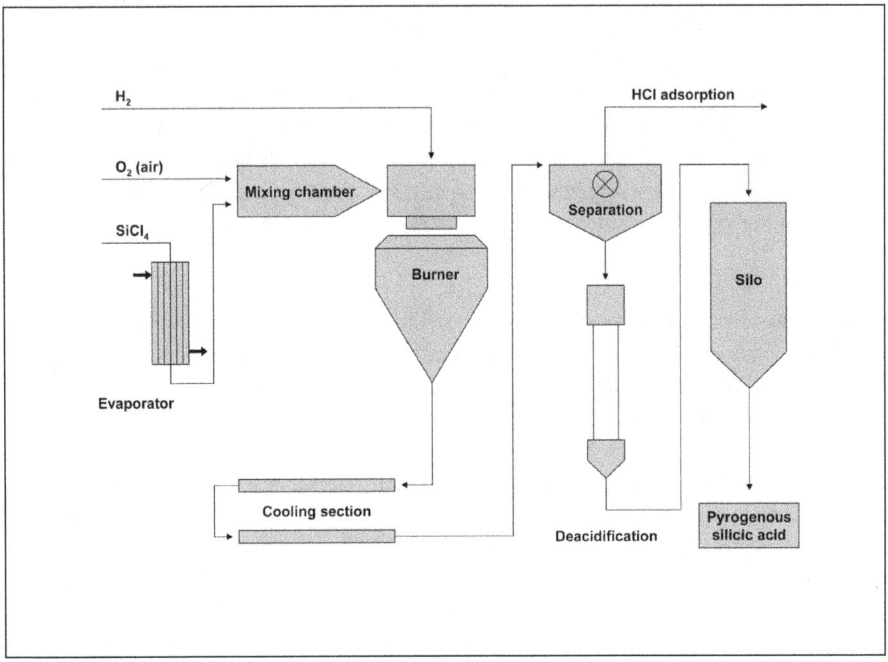

Figure 3.23: Schematic production flow for pyrogenous silicic acid

3.2.4.1 Pyrogenous silicic acids

Pyrogenous silicic acid, or fumed silica, is a general term used for highly dispersed silicic acids (HDK) [8-10]. It is produced on an industrial scale by flame hydrolysis (the Aerosil process). This works by spraying volatilised silicon tetrachloride into a hydrogen and oxygen detonating gas flame at a temperature of 1000 °C. Temporarily formed water reacts spontaneously to produce highly dispersed silicic acid.

$$H_2 + O_2 + SiCl_4 \xrightarrow{1000\ ^\circ C} SiO_2 + 4HCl$$

Equation 3.10: Decomposition of silicon tetrachloride to form highly dispersed silicic acid

Changes in flame temperature, reagent concentration, and how long the silicic acid is exposed in the combustion chamber all strongly influence the reaction. Accordingly, silicic acid can be produced with highly controllable particle size, particle size distribution, specific surface area, and morphology. Gaseous hydrogen chloride forms as a by-product, and is separated from the silicic acid. The obtained surface of the pyrogenous silicic acid particles is almost without pores. With suitably adjusted reaction conditions, alternatives to silicon tetrachloride are silane, trichlorosilane, and methyltrichlorosilane.

Silanol groups lend synthetic silicic acids a hydrophilic character. Accordingly, they are frequently given chemical after-treatment, with chlorosilane for example, which reacts with the hydroxyl groups. This produces the desired hydrophobic surface properties, extending the possible application spectrum. ISO 3262 part 20 distinguishes pyrogenous silicic acids into types A (hydrophilic, untreated) and B (hydrophobic, after-treated), as well as defining high purity of at least 99.8 %.

3.2.4.2 Precipitated silicic acid

Precipitated silicic acid represents the bulk of silicic acid production. Precipitation occurs in a dilute alkali silicate solution such as sodium silicate, with the admixture of sulphuric or some other mineral acid. By continuously feeding chemical stock to the precipitation reactor under constant alkaline conditions, silicic acid precipitates as colloidal, primary particles. Key process parameters for controlling the reaction are temperature, concentration of the solution, and convection. Marking the end point of the precipitation reaction is a shift from alkaline to acid pH, which stabilises the precipitate. Subsequent filtration and washing of the silicic acid reduces undesired by-products like sodium sulphate to <1 %. The purification process is followed by drying, which can be done in rotary, flash, or spray dryers for example. Drying almost always creates agglomerates of widely differing size and distribution, depending on the dryer used. Conventional drying methods therefore necessitate an additional grinding stage.

Precipitated silicic acids contain around 98 % silica, above the threshold figure of 95 % specified in ISO 3262 part 19. The standard also distinguishes types A and B, which differ mainly in their hydrophobic properties as the result of organic surface coating: after-treatment to produce a hydrophobic surface is beneficial for precipitated silicic acid in certain applications. Dimethyldichlorosilane, silicone oils and waxes may be used for coating; the choice ultimately defines the product's application areas.

3.2.5 Precipitated aluminium silicate

Precipitated aluminium silicate is produced in basically the same way as precipitated silicic acids. It is nonetheless a special case, because precipitated aluminium silicate is an "artificially contaminated" precipitated silicic acid. The term "synthetic aluminium silicate" is thus chemically incorrect; it should properly be "precipitated sodium aluminium silicate". This term does not seem to have caught on amongst users of the product, though. Just like regular silicic acid production, the starting chemicals sodium waterglass and sulphuric acid are necessary for the precipitation reaction. But then a third component, aluminium sulphate, is added. Precipitation proceeds in essentially the same manner as for regular precipitated silicic acids (see Chapter 3.2.4.2), but produces a different mix of compounds.

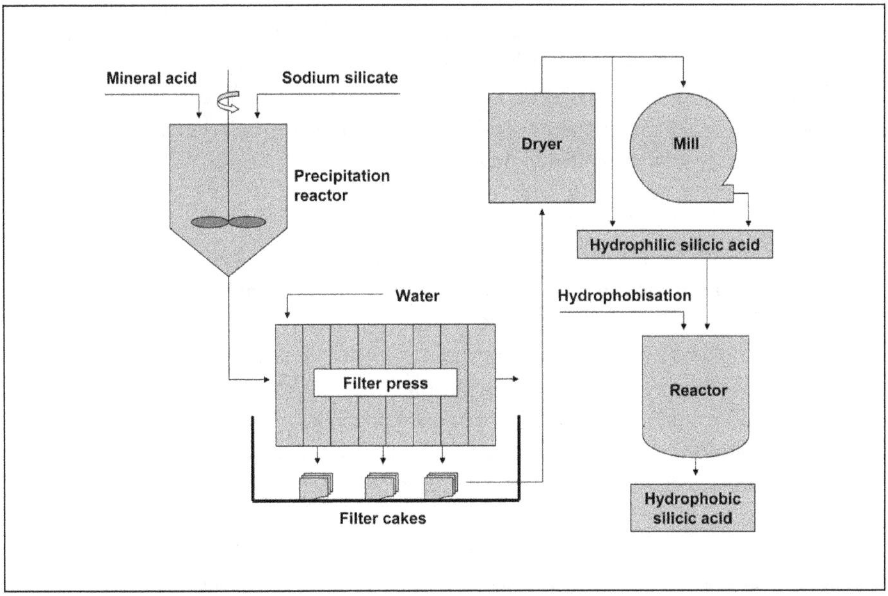

Figure 3.24: Generalised schematic production flow for precipitated silicic acid

This is described in ISO 3262 part 18, which states a minimum SiO_2 content of >75 %, Na_2O content of 3 to 10 %, and Al_2O_3 content of 5 to 15 %. Commercial products typically contain 8 % sodium oxide, 10 % alumina (the "contamination"), and 82 % silica.

3.3 Surface treatment of fillers

Fillers exhibit characteristic surface properties according to their manufacture. Their surfaces depend on the specific material, but are mostly hydrophilic. Yet a hydrophilic surface can be disadvantageous or inadequate in some applications. Hence the option of coating the filler surface, either in the final stage of production, or as a later after-treatment. The (usually organic) coating material encases the filler particles with a monomolecular or thicker layer. Most inorganic coatings are used only in pigments. Coating can be done either in aqueous suspension of the filler or on the powder itself. Various materials in differing states are available. They adhere to the filler surface by chemisorption or physisorption, depending on coating conditions. Nevertheless, chemical analysis will usually reveal a certain amount of free fatty acid that has not adhered by either mechanism. In other words, a 100 % combination of the fatty acid with the filler surface is unattainable in practice. Although figures of 90 % and higher are not unknown, the extent of conversion is often just 50 to 70 %. Coating quality can be assessed by various methods. Among the indirect approaches are before and after comparison of the decrease in tamped and bulk volume, or absorption of linseed oil or vaseline oil (VO). The proportion of unconverted fatty acid can be determined analytically, by titration. Stearic acid is the prevalent coating for carbonate, silica and silicate-based fillers. However, stearic acid does not exist as a single, isolated fatty acid, but actually contains oleic, palmitic, myristic and lauric acid. Equation 3.11 illustrates coating with stearic acid.

$$CaCO_3 + 2\,C_{17}H_{35}COOH \xrightarrow{\text{min. } 80\ ^\circ C} Ca(C_{17}H_{35}COO)_2 + CO_2 + H_2O$$

Equation 3.11: Conversion reaction of stearic acid with calcium carbonate

Adding thermal energy would ideally convert 100 % of the stearic acid on the surface of the calcium carbonate particles to calcium stearate. Yet as described previously, conversion is incomplete. The coating amount [11, 12] generally depends on the specific surface area of the filler. A rule of thumb is 0.25 % by weight of coating material per m^2 of surface. The maximum coating is commonly around 3 to 3.5 %, even when the specific surface area is greater than 12 to 14 m^2/g.

The surface of fillers like pyrogenous silicic acids can also be coated with organofunctional silane (amino-, epoxy-, vinylsilane and others), siloxane, silicane, or titanate [8, 10]. These coatings have ester bonds that hydrolyse in the presence of water, while the silane or titanium forms chemical or physical bonds with hydroxyl groups on the filler surface. The filler has been effectively functionalised in terms of its ability to bond with other ingredients in a formula.

$$2R-Si-OH+2(CH_3)_2SiCl_2 \xrightarrow{H_2O} R-Si-O-Si(CH_3)_2-O-(CH_3)_2Si-O-Si-R+4HCl$$

Equation 3.12: Conversion reaction of a silicic acid surface with two molecules of dimethyldichlorosilane

Coating often improves dispersal behaviour in non-polar systems. A strongly hydrophobic surface character makes the filler even more functional, for more water-resistant, hydrophobic coatings (outdoor applications). Compared with an uncoated product, the filler itself absorbs less moisture enhancing the storage stability of non-aqueous systems. Silane based surface coatings bear positively on the way fillers adhere to substrates in conjunction with binders. All the same, coated fillers represent just a very small proportion of annual filler production, and their deployment in paints and coatings is lesser still. Most find their way into plastics, adhesives and sealants.

3.4 References

[1] Tegethoff, F. W., Calcium Carbonate – From the Cretaceous Period into the 21st Century, p. 171–191, Birkhäuser Verlag, Basel, 2001

[2] CCA-Europe – The European Calcium Carbonate Association, Internet www.ima-eu.org/en/ccaprod.html, 2004

[3] Socal documentation, "Precipitated Calcium Carbonate for Paints", Solvay, Rheinberg, 2001

[4] Kittel, H., "Lehrbuch der Lacke und Beschichtungen – Pigmente, Füllstoffe und Farbmetrik", vol. 5, p. 447–448, S. Hirzel Verlag, Stuttgart, 2003

[5] Tegethoff, F. W., Calcium Carbonate – From the Cretaceous Period into the 21st Century, p. 191–192, Birkhäuser Verlag, Basel, 2001

[6] Wypych, G., "Handbook of Fillers", p. 48–57, ChemTec Publishing, Toronto, 1999

[7] Hollemann-Wiberg, "Lehrbuch der anorganischen Chemie", vol. 101., p. 922–950, Gruyter, Berlin 1995

[8] Degussa documentation, "Grundlagen von Aerosil", Fine Particles series of papers, number 11, issue no. 7

[9] DBP 762723, Degussa AG, 1942

[10] Wacker Documentation, "Herstellung von Wacker HDK", Internet www.wacker.de, 2005

[11] Kovacevic, V., Lucic, S., Hace, D., Cerovecki, Z., Adhesion Science Technology, vol. 10, p. 1273–1285, no. 12, 1996

[12] Balard, H., Papirer, E., "Characterization and modification of fillers for paints and coatings", Progress in Organic Coatings, vol. 22, no. 1–4, p. 1–17, 1993

4 Characterisation of fillers

Quality requirements for basic raw materials are comparable to those of the end products that are made from them. To meet these requirements, specific filler properties require regular monitoring. Monitoring delivers data on product consistency, which in turn provides insight into the manufacturing process. This is particularly important with fillers, on account of the many natural raw materials involved. They can all vary in their composition, even within the same quarry, and this is especially true of mineral fillers. That is why prospecting (see Chapter 3.1.1) is such an important activity: it seeks to identify deposits of very high quality that are extremely homogenous as well. Characterisation of fillers that will be produced from the deposit actually begins at this stage, where analytical techniques are applied to determine chemical and mineralogical composition.

Those same techniques are also used for quality control in the course of mining activity. After initial quality checks further tests come to assess what properties can be expected of the mineral filler at the end of the production chain. The filler manufacturer tests his products in-house, and the subsequent buyer does the same with incoming goods. Such testing will typically focus on properties like filler brightness, and particle size distribution.

Ex post facto identification of a filler is another frequent issue. This is a case for filler analysis, using preparation and analysis techniques of varying complexity that depend on the sample material. Analysing a filler, or blend of fillers, contained in already-formulated paints and coatings is especially demanding in terms of delivering qualitative and quantitative answers. Today's analytical instruments can measure traces on the order of just a few particles per million, or even billion, within just a few minutes. That calls for greater care and diligence than ever in sample preparation, which is actually more time-consuming than actual measurement and interpretation of the results.

4.1 Filler testing

Numerous tests may be applied to characterise fillers. Many are based on national or international standards: the Paints and Varnishes Handbook published by the International Organisation for Standardisation (ISO) lists a multitude of raw materials tests [1]. The pre-eminent and most comprehensive standards for fillers are ISO 787 parts 1-25 for tests, and ISO 3262 parts 1-21 for specifications. These standards form the basis on which producers test their fillers before they leave the factory, and users of those products are likely to refer to them as well for quality checks on incoming goods. The same standards may

Table 4.1: Summary of filler testing methods

Testing method	Remarks	Test conditions
Density	Pycnometer method	ISO 787-10
Volatile contents at 105°C		ISO 787-2
Hydrochloric acid-insoluble content		
Water-soluble content	Hot extraction method Cold extraction method	ISO 787-3 ISO 787-8
Yellowness index		DIN 6167
Brightness		DIN 53163
Oil absorption		ISO 787-5
Particle size analysis	Sieve analysis	DIN 66165
Particle size distribution	Sedimentation analysis	DIN 66115
pH value	Aqueous suspension	ISO 787-9
Bulk density/volume		ISO 903
Sieve residue	Selectable mesh size	ISO 787-7 ISO 3310
Specific surface area	BET method	ISO 9277
Specific electrical resistivity	Aqueous extract	ISO 787-14
Tamped density/volume		ISO 787-11
Whiteness		ISO 2469 ISO 2470

also be a factor in ISO certification, such as ISO 9001-9004 for filler production and ISO 14001 for mining minerals. Other standards relating to fillers, like ISO 8780, cover testing of application-related properties.

4.1.1 Optical properties

Optical properties are of paramount importance in fillers intended for universal use – which are preferably very bright, and neutral in colour. A material's whiteness or brightness is a significant factor of subjective colour impression. The only way to objectively evaluate that colour impression is to measure under defined conditions of technique, geometry, and illumination [2]. Controlled measurement provides comparable figures for objective colour appraisal, as opposed to a subjective impression. It should be noted that geometry is a fixed characteristic of the spectral photometer used: d/0 (widespread in the

Table 4.2: Whiteness determination and calculation parameters

Colour value	Standard light type	Geometry	Standard
R_x, R_y, R_z	C	2°	DIN 53145 (DIN 53140, 53163)
R_{457}	D65	10°	ISO 2469
L*	D65	10°	DIN 6174
Yellowness index	D65	10°	DIN 6167

filler and paper industries), or d/8 (common in the paints and coatings industry). However, it possible to pre-select a standard illumination type and set the instrument accordingly. The spectral photometer accounts for this, along with the "geometry", in its internal conversion of measured values to whiteness or brightness figures (see Table 4.2).

Published literature tends to use the terms "whiteness" and "brightness" interchangeably where fillers are concerned. That makes it especially important to state not only the measurement figures, but also a clear reference to the applicable standard. The underlying measurement technique is always basically the same: reflectivity over the visible light spectrum, referenced to a barium sulphate white standard defined by DIN 5033. The resulting spectral curves are used to calculate standard colour values X, Y and Z. The value Y additionally represents brightness, scaled along an axis vertically perpendicular to the achromatic point of the two colour planes. The CIE-L*a*b* system (DIN 6174) is one of several colour systems in widespread use. It expresses the chromatic components in terms of a* and b*, plus brightness L*. The standard colour values X, Y and Z also provide a basis for calculating other colour-related factors, like whiteness. A whiteness value alone is not enough, though, because its exact significance depends on the calculation standard used (DIN 53140, DIN 53145, DIN 53163, ISO 2469, ISO 2470, etc.), see Chapter 4.1.1.1. Overall, it is worth remembering that given a choice of brightness, and whiteness expressed, for example, as R_y (reflectivity factor under a green filter), whiteness is the fairer measure because it is better at differentiating between fillers.

4.1.1.1 Whiteness

The principles of whiteness have already been discussed in Chapter 4.1.1.1. Precise measurement of whiteness involves spectral photometer methods. Preparing samples of powdered products is a particularly important step, and uses a special press to make powder tablets out of the filler under measurement. Pressing force and sample quantity depend on the mineral concerned, and require once-only determination by preliminary trials. Calibration references a DIN 5033 barium sulphate white standard. Actual measurements

Table 4.3: Comparative whiteness and brightness for some typical fillers

Filler	R_y whiteness	R_{457} whiteness	L* brightness
GCC A	83.6	78.1	93.3
GCC F	94.9	93.5	98.0
Dolomite A	93.3	92.9	97.4
Kaolin A	84.7	79.9	93.7
Talcum A	90.9	89.8	96.3

determine reflectance values for the filler across the spectrum from 400 (360) to 700 nm. Colour values X, Y and Z are then calculated from the resulting spectral curve values. Whiteness values or reflectance factors are obtainable by further conversion, which modern measuring instruments accomplish automatically. Table 4.2 lists a summary of various whiteness specifications and their associated parameters.

Datasheets on filler products for the paints and coatings industry customarily specify whiteness as R_y, i.e. reflectance values under a green filter. R_{457} whiteness, also known as Tappi whiteness, is mainly prevalent in the paper industry. In practice, R_y whiteness values work out slightly higher than their R_{457} counterparts: see the examples in Table 4.3. Whiteness purity is specified by a supplemental yellowness index according to DIN 6167, see Chapter 4.1.1.3.

4.1.1.2 Brightness

Brightness is generally understood as the strength of a sensation of light. Brightness of fillers according to CIE-L*a*b* is determined by the same measuring principle used for whiteness, with certain differences in the type of illumination and the geometry (see Table 4.2), and the way X, Y and Z colour values are transformed to L* brightness.

Table 4.3 clearly illustrates how whiteness values differ considerably more than the corresponding L* brightness. L* brightness is not in widespread use, given the superior differentiation obtainable by specifying R_y whiteness. Nevertheless, L* brightness is often specified as well because of the simple and understandable arrangement of the CIE-L*a*b* colour system, and its popularity in the paints and coatings industry.

4.1.1.3 Yellowness index

In specifying filler whiteness, it is also useful to state yellowness, because whiteness by itself is an insufficient qualitative statement. Yellowness index is described in DIN 6167, which amongst other things sets out how X, Y and Z colour values are transformed via R_x, R_y and R_z into yellowness index.

*Yellowness index = 100 * [(R_x-R_z)/R_y]*

Analogous to the previous comparison of R_y vs. L* brightness, measurements of yellowness index are more meaningful than the corresponding b* values. Table 4.4 shows a side-by-side comparison of results for five different mineral fillers: their differences in yellowness index are considerably greater, which means better resolution and hence precision than is possible with the corresponding b* values.

Table 4.4: Comparative yellowness index for some typical fillers

Filler	Yellowness index	b*
GCC A	9.2	+4.62
GCC F	1.9	+0.99
Dolomite A	0.6	+0.32
Kaolin A	7.2	+3.87
Talcum A	1.5	+0.85

4.1.1.4 Refractive index

Refractive index is an important material characteristic. A beam of light changes its direction of travel as it passes from one medium to another at a non-perpendicular angle to their boundary. The amount of change depends on the refractive indices of the respective media and finds general expression in Snell's law of refraction [3], shown in Equation 4.1.

$$\frac{\sin\theta_1}{\sin\theta_2} = \frac{v_1/c}{v_2/c} = \frac{n_2}{n_1} = n_{21}$$

Equation 4.1: Snell's law of refraction

The relative refractive index n_{21} is the ratio of the absolute refractive index of medium 2 (n_2) vs. medium 1 (n_1). The same can also be expressed in terms of the angles of the incident light beam θ_1 and the refracted beam θ_2, or the speed of light $v_{(1,2)}$ in the media versus in a vacuum, c.

Measuring refractive index n in solid materials like minerals is more difficult than with liquids. Measurement is performed on a polished mineral surface that is placed on the refractometer with the aid of contact liquid. The smaller the inter-atomic distances within the material, the higher its refractive index n. Optically anisotropic crystalline materials have the special property of being able to polarise light, i.e. incident light becomes split into differing spatial planes (a phenomenon known as double refraction).

Fillers exhibit similar refractive indices to organic materials like styrol-acrylate dispersions, alkyd resins, etc. For this reason, fillers do not generally contribute to opacity in paint systems. The exceptions are paint systems formulated above the critical pigment volume concentration (CPVC): here, the coating film includes air pores in addition to mineral fillers

Table 4.5: Refractive indices for various materials

Material	**Refractive index/range**
Air	1.00
Water	1.33
Fluorspar	1.43
Glycerine	1.45
Quartz modifications	1.49 to 1.55
Calcite	1.49 to 1.66
Orthoclase	1.52 to 1.54
Phenol	1.54
Spinelle	1.72 to 1.75
Titanite	1.89 to 2.05
Diamond	2.42
Rutile	2.62 to 2.90

and organic binders. Air has the lowest refractive index n of around 1.0, which is sufficiently different from that of the filler or organic binder for strong light refraction to occur and hence contribute to the opacity of the dry paint film; see also Chapters 6.2.3, 6.2.4, 6.3.3 and 6.4.1.

4.1.2 Morphology

Not only do mineral fillers exist in multiple crystalline forms, their morphology can differ too – mainly in terms of particle size, particle shape, the structure of the particles, and their surface properties. Qualitative determination methods can involve modest to highly complex instruments. This chapter examines methods for determining particle size distribution, specific surface area, and oil absorption. Scanning electron microscopy makes an excellent method for further examination of particle morphology; see Chapter 4.2.1.

4.1.2.1 Particle size distribution

Particle size distribution is a defined means of stating particle diameter in substances like powders, dusts and granulates. It is expressed in terms of volume or percentage mass within a defined range of particle sizes. Determination of particle size distribution is one of the primary methods for characterising fillers. Various techniques exist [4], and the choice of which one to use will generally hinge on the expected fineness of the filler – no single measurement technique can claim to be universal, although some of them do cover a broad spectrum of particle sizes. The choices are:

- Sieve analysis (ISO 787 part 7)
- Sedimentation analysis (DIN 66115 parts 1+2)
- Laser diffraction analysis (ISO 13323 parts 1+2).
- Optical measurement (Camsizer)

Each one is an approach to measuring irregularly shaped bodies, which entails first measuring the bodies' volume, then stating it as the diameter of a sphere with identical volume (equivalent diameter). Figures for the median particle size and the coarsest particle size (top cut) are important criteria for users of fillers, and both can be read-off from the particle size distribution curve. D50%, the median particle diameter, is where the curve intersects 50 % by weight of all particles with a given size. The choices of 98 % or 95 % by weight for the top cut intersect rather than 100 % are made on grounds of reproducibility. In the example shown in Figure 4.1, the median particle diameter D50% is 0.75 µm and the top cut D98% is 4 µm.

Sieve analysis is used today with very coarse particles like granular fillers. A sieving machine uses stacked sieves of differing coarseness to measure particle size distribution by the method known as tower sieving. Differently sized particles are separated into fractions that are quantified by weighing back. As well as particle size distribution, sieving can determine the sieve residue (screen oversize) of filler for a given mesh spacing, which features frequently in filler specifications. This can be done by hand-sieving (dry or wet), air-jet sieving or ultrasonic sieving. The mesh used for sieve analysis can go down as far as 10 µm, with 45 µm being a common figure. There is practically no upper limit to the particle sizes that can be accommodated using this method. In Anglo-Saxon countries, the term 325 mesh is frequently used (= 325 mesh apertures per square inch, all identically sized), which corresponds to a mesh pitch of 45 µm; see also Figure 4.2.

The sedimentation method of determining particle size distribution is based on the way particles suspended in a liquid settle according to their respective weights. The principle is expressed by Stokes' law, see Equation 4.2.

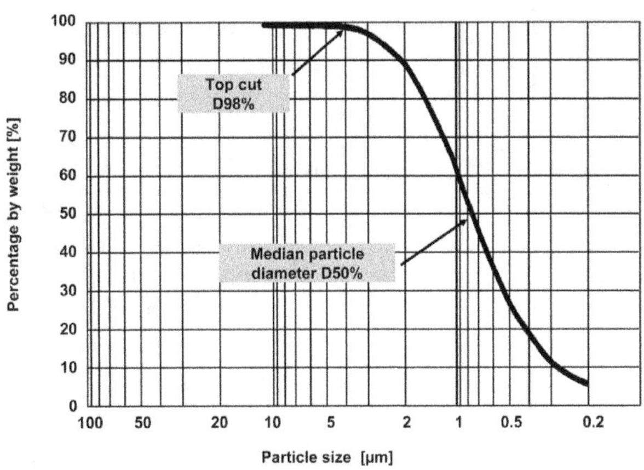

Figure 4.1: Filler particle size distribution showing the median particle diameter D50% and the top cut D98%

$$v = \frac{2}{9\eta} \cdot (\rho_s - \rho_l) g r^2$$

Equation 4.2: Stokes' law

The settling, or sedimentation rate v depends on particle radius r, material density ρs, measuring fluid density ρl (constant, as a rule), and the effect of gravity, which is why the method cannot be used on mixtures of more than one material. The rate is measured by sampling particle concentration and size at a defined height, by X-ray absorption for example. Instruments based on this principle cover a measuring range of 0.1 to 50 µm.

Laser diffraction analysis is based on the principle of light scattering (with optional back-scattering) from tiny particles in emulsion or suspension according to the Fraunhofer or Mie theory. The illumination source is a He-Ne laser. Instruments that deploy this method cover a relatively broad spectrum of particle sizes from 0.05 to 900 µm, or 0.7 to 400 µm.

The methods described above are nevertheless unsuited to determining particle size distributions in very fine, precipitated fillers like PCC and silicic acid. The fineness of such materials is determined in terms of the median particle diameter only, using scanning electron microscopy (see Chapter 4.2.1) or by measuring air permeability of a pressed filler tablet by the Blaine method (determination of specific surface area, Chapter 4.1.2.2).

Recent engineering developments led to instruments performing particle counting, size and shape analysis in real time and high resolution based on the principle of digital image processing. The sample is fed via a vibrator chute to the measuring field, where dozens of millions measuring points are recorded per second, see also Figure 4.5. By the means of process orientated software different analyses can be calculated. The detectable size of particles ranges from some ten microns to some ten millimetres. Therefore, this method is a fast and efficient alternative to the before described sieve analysis with full compatibility. In addition, it delivers information about the length, diameter, width/length aspect ratio, roundness, symmetry and convexity of particles. Besides the scien-

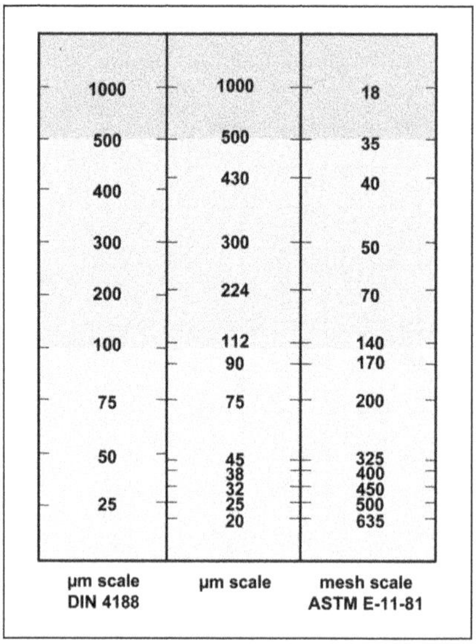

Figure 4.2: Micrometer vs. mesh scale particle size comparison

OMYA AG, CH-4665 OFTRINGEN

SediGraph III V1.00 | Unit 1 | Serial Number: 104 | Page 1

Sample: SIKRON SH 300 04.01.2005 05 8050
Operator: H86Q01 pv
Submitter: GQM-GROUP QUALITY MANAGEMENT
File: F:\...\PC230CPY\002-499.SMP
Material/Liquid: Quarz / Wasser+0.1%Na4P2O7
Measurement Principle: X-Ray monitored gravity sedimentation
Calculation Method: Stokes sedimentation and Beer's law of extinction

Test Number: 1
Analyzed: 25.01.2005 19:31:07
Reported: 02.03.2005 15:10:04
Liquid Visc: 0.7897 mPa·s
Analysis Temp: 30.7 °C
Full Scale Mass: 100.0 %

Analysis Type: High Speed(Adj)
Run Time: 0:29 hrs:min
Sample Density: 2.650 g/cm³
Liquid Density: 0.9955 g/cm³
Base/Full Scale: 136 / 93 kCnts/s
Reynolds Number: 0.18

Combined Report

Report by Size Table

Low Diameter (µm)	Cumulative Mass Finer (Percent)	Low Diameter (µm)	Cumulative Mass Finer (Percent)	Low Diameter (µm)	Cumulative Mass Finer (Percent)	Low Diameter (µm)	Cumulative Mass Finer (Percent)
50.00	98.7	10.00	61.0	2.000	10.9	0.400	2.1
40.00	99.4	8.000	49.7	1.500	8.1	0.300	1.9
30.00	99.3	6.000	37.5	1.000	4.8	0.200	0.3
25.00	97.7	5.000	31.3	0.800	3.5		
20.00	92.6	4.000	24.8	0.600	2.4		
15.00	81.4	3.000	18.0	0.500	2.3		

Report by Mass Percent

Low Diameter (µm)	Cumulative Mass Finer (Percent)	Low Diameter (µm)	Cumulative Mass Finer (Percent)	Low Diameter (µm)	Cumulative Mass Finer (Percent)
25.54	98.0	8.043	50.0	3.285	20.0

Cumulative Finer Mass Percent vs. Diameter

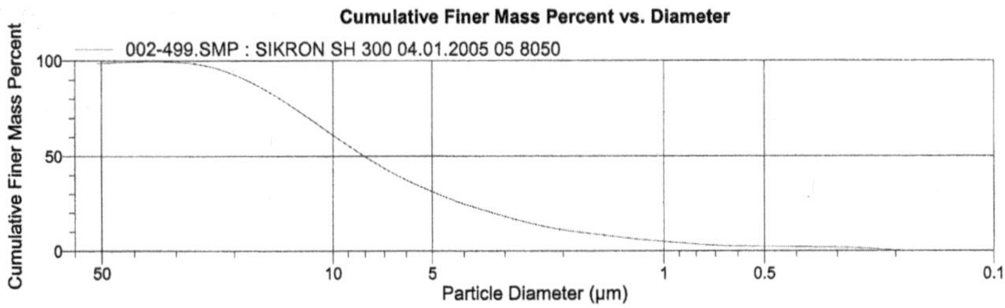

Figure 4.3: Printout of particle size distribution results for quartz B, obtained by sedimentation analysis (Sedigraph 5100)
source: Omya AG

tific particle analyses the used process allows a high sample throughput, which can be further increased with an auto sampler, qualifying the instrument for continuous quality monitoring plus measuring protocol based on DIN 66165.

CILAS

PARTICLE SIZE DISTRIBUTION

CILAS 920 Liquid

Range : 0.30 mu - 400.00 mu / 30 Classes

Sample Ref	: 05 8050 SIKRON SH 300	Ultrasounds	: 60 s (+during)
Type produit	: QUARZ	Concentration	: 238
Client	:	Diameter at 20%	: 2.80 mu
Comments	: H9MR02	Diameter at 50%	: 9.07 mu
Liquid	: Wasser	Diameter at 98%	: 33.66 mu
Dispersing agent	: Natriumpyrophosphat	Fraunhofer	
Operator	: BH	Density/Factor	---------
Company	: OMYA AG	Specific surface	---------
Location	: CH-4665 Oftringen	Auto. dilution/top up	: Yes / Yes
Date : 19.01.2005	Time : 10:15:28	Nb Measur./Rins.	: 20 / 3
Index meas.	: 10097		

Customer defined classes — in volume / undersize

x	0.70	1.00	2.00	4.00	5.00	8.00	10.00	12.00	20.00	25.00
Q3	4.19	6.17	13.67	27.88	33.19	45.96	53.15	59.59	80.61	90.10

x	32.00	45.00	71.00	100.0	125.0	160.0	200.0	250.0	315.0	400.0
Q3	97.24	100.00	100.00	100.00	100.00	100.00	100.00	100.00	100.00	100.00

x : diameter / mu Q3 : cumulative value / % q3 : population density / %

User percentages

%	20.00	50.00	98.00
Q3	2.80	9.07	33.66

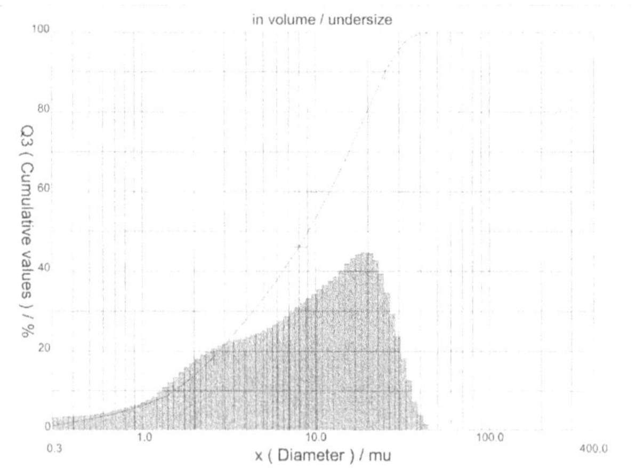

Figure 4.4: Printout of particle size distribution results for quartz B, obtained by laser diffraction analysis (Cilas 920) *source: Omya AG*

4.1.2.2 Specific surface area

Filler specific surface area is defined as the mass- or volume-related surface area. ISO 9277 defines this as the entire surface, inclusive of internal cavities. The filler industry predominantly uses the BET method, named after Brunauer, Emmett and Teller [5]. Determining the Brunauer, Emmett and Teller adsorption isotherm is the basis for calculating surface area and pore size. At a temperature of -196 °C, the entire sample surface has adsorbed a monomolecular layer of gaseous nitrogen. The specific surface area in m^2/g is calculated from the amount of adsorbed gas and the quantity of sample material.

The Blaine airflow method (DIN 66126) is another means of determining specific surface area. In contrast to the BET method, specific surface area is stated here in cm^2/g.

The finer a filler is, the greater its specific surface area. Next to fineness, particle shape also has a crucial effect on specific surface area. Thus, lamellar (platelet-shaped) fillers have greater specific surface areas than nodular fillers of comparable particle size. Porous fillers like kieselguhr can sometimes have a considerable internal surface area: in Figure 4.6, very coarse kieselguhr A has a specific surface area similar to the fine nodular fillers also shown.

Figure 4.5: Analysis of particle size distribution for granules with a Camsizer source: Omya AG

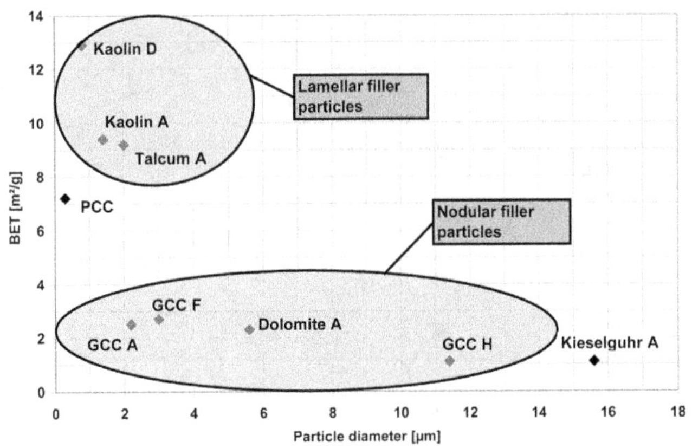

Figure 4.6: Correlation of filler particle size and shape with BET specific surface area in a variety of fillers

4.1.2.3 Oil absorption

Determination of oil absorption according to ISO 787 part 5 is a classical test for fillers. Oil absorption describes a filler's oil demand (varnish linseed oil). Stated more precisely, the method describes a means of reproducibly determining the amount of varnish linseed oil required to completely wet 100 g of filler. The instruments required are modest: varnish linseed oil is dispensed from a burette in dosed increments and mixed with the filler on a glass plate, using a spatula and moderate pressure. Oil is added until the mix forms a paste that is cohesive but not yet sticky, and readily spreads into a thin film. The endpoint – plainly recognisable with most fillers – is when scraping the paste away with the spatula produces cohesive tiny rolls (like miniature butter curls). All the same, reproducibility can deviate in practice, especially when separate, and perhaps inexperienced testers perform the measurement.

Oil absorption by fillers, like their specific surface area, goes up with increasing fineness. As a rule, oil absorption largely correlates with the specific surface area. The exceptions are fillers whose inner surfaces cannot be wetted with varnish linseed oil, or include very large cavities that do not show up in the determination of specific surface area. Figure 4.7 clearly shows how nodular fillers have considerably lower oil absorption compared to lamellar fillers.

4.1.3 Physical properties

Physical properties like density, bulk and tamped density, electrical conductivity and Mohs hardness can be quite influential on product properties, logistics and economics. For ex-

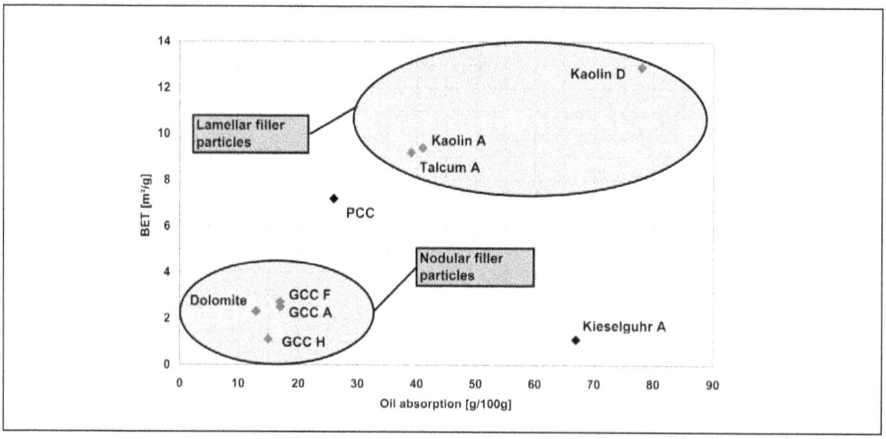

Figure 4.7: Correlation of BET specific surface area with oil absorption in a variety of fillers

ample, using lower-density fillers in paints and coatings can bring economic benefits for high-volume sales. Filler bulk and tamped density, and electrical conductivity, are crucial to ease of handling when material is packaged for transportation, shipped, stored and incorporated into paint and coating systems. The abrasiveness of paints and coatings affects their processing and application; to a large extent this is a function of mineral hardness characterised by the Mohs scale. Density, electrical characteristics, and Mohs hardness are material constants, whereas bulk and tamped density may vary depending on processing.

Table 4.6: Densities of some typical fillers

Filler	**Density [g/cm³]**
Light weight fillers	< 0.2
Kieselguhr	1.9 to 2.65
Precipitated silicic acid	1.9 to 2.3
Pyrogenous silicic acid	2.2
Cristobalite	2.35
Kaolin	2.3 to 2.6
Quartz	2.5 to 2.7
Calcium carbonate	2.7
Talcum	2.7 to 2.9
Mica	2.7 to 2.9
Dolomite	2.8 to 2.9
Barium sulphate	4.0 to 4.4

4.1.3.1 Density

By definition, the density of a uniform substance is its mass per unit volume, i.e. the mass in g or kg contained in 1 cm³ or 1 l.

$$Density = \frac{Mass}{Volume}\left[\frac{g}{cm^3}\right]$$

Equation 4.3: Calculation of density from the mass and volume of a substance

Density of solid materials is usually determined using a Gay-Lussac pycnometer. The principle is based on displacement of measuring fluid by a solid material. The exact measurement method for fillers is defined in ISO 787 part 10 and may be performed by method A (without vacuum) or method B (with vacuum).

4.1.3.2 Bulk density

In addition to filler materials' inherent density, their bulk density is highly relevant to formulating paint and coating systems. Bulk density is the quotient of mass and the volume occupied by that mass, which includes spaces between particles as well as internal cavities. In practical terms, this is the volume occupied by the filler in its loose state. Measurement of bulk density is standardised in ISO 697, and is determined indirectly via bulk volume. For measurement purposes, 100 g of filler is loosely filled into a suitable measuring vessel, e.g. a measuring cylinder, where the bulk volume is read off. Bulk density is then calculated according to Equation 4.4.

$$Bulk\ density = \frac{Mass}{Bulk\ volume}\left[\frac{g}{ml}\right]$$

Equation 4.4: Calculation of bulk density via bulk volume

Fillers have considerably lower bulk density than their inherent density. Bulk density is needed to calculate the size of transportation packaging, e.g. silo trucks, big bags and sacks.

4.1.3.3 Tamped density

The third key specifier of density is tamped density. Like bulk density, tamped density is determined via volumetric measurement, see ISO 787 part 11. Tamped volume is the smallest volume occupied by a given quantity of filler that has been compacted under standardised conditions. Reproducible determination of tamped volume is ensured by 1,250 tamping cycles on 100 g of filler in a measuring cylinder, using a tamping volumeter. The procedure is repeated until successive volume determinations differ by less than 2 ml. From knowledge of the tamped volume, it is possible to calculate tamped density, shown by Equation 4.5.

$$Tamped\ density = \frac{Mass}{Tamped\ volume}\left[\frac{g}{ml}\right]$$

Equation 4.5: Calculation of tamped density via tamped volume

Once a filler's tamped density has been determined, conclusions can be drawn about conditions in filler silos, especially near the bottom. For producers as well as users of fillers, it is therefore an important parameter for structural calculations and expected loads in silo designs.

4.1.3.4 Volatile matter

Powdered fillers can contain various kinds of volatile matter like moisture, and organic compounds from milling agents or surface coatings. Actual measurement attempts to determine the quantity of volatile matter, as distinguished by tests at differing temperatures.

- Volatile matter at 105 °C according to ISO 787 part 2
- Volatile matter at 105 to 400 °C according to ISO 787 part 2

Determination of volatile matter at 105 °C primarily ascertains the amount of moisture adsorbed by the filler surface. The amount increases along with specific surface area, and may be anything up to 1 % depending on the filler's fineness and hydrophilic properties. Certain fillers contain bound crystal water that may dissociate below 105 °C, depending on the material. However, the temperature often has to be higher than 105 °C before dissociation occurs.

Determination of organic volatile matter content like milling agents and surface coatings is done in a muffle furnace at 400 °C. Thermogravimetric analysis (TGA), see Chapter 4.2.4.1, can provide a more precise picture about dissociation of volatile matter, like the temperatures and energies involved.

4.1.3.5 Mohs hardness

Minerals are of differing hardness, depending on their specific crystalline structure. Mohs devised a simple, unitless scale from 1 to 10 for characterising mineral hardness. Classification within this scale is defined such that a mineral with a higher degree of hardness is just capable of injuring the surface of a mineral having one degree less hardness, for example by scoring. Accordingly, a Mohs hardness of 1 denotes the least, and 10 the greatest mineral hardness. Table 4.7 lists a variety of minerals and their hardness on the Mohs scale.

Table 4.7: Mohs scale for classification of minerals by hardness

Mohs hardness	Filler
1	Talcum
2	Gypsum
3	Calcium carbonate
4	Fluorspar
5	Apatite
6	Orthoclase
7	Quartz
8	Topaz
9	Corundum
10	Diamond

Mohs hardness is a good way to deduce the behaviour of mineral fillers. Minerals with a hardness of 1, like talcum, create virtually no abrasion in processing and produce coating films that are easy to sand and smooth. Minerals with Mohs hardness of 3 or 4 also produce little abrasion, which makes them most effective in formulations for very bright and neutral colour shades. Although very hard minerals like quartz produce heavier abrasion in emulsification and application equipment, their Mohs hardness of 7 means they are ideal in formulations for highly wear-resistant coating films.

4.1.4 Chemical properties

Certain filler applications require special chemical resistance and properties. Accordingly, tests for chemical properties like pH, water-soluble matter and hydrophobicity feature frequently in filler characterisation. Knowledge of these properties provides a selection basis for fillers used in paint and coating systems, especially for outdoor applications with specialised corrosion protection needs.

4.1.4.1 pH value

It is essential to know the pH of fillers for use in aqueous systems: selecting a filler with the wrong pH can be fatal to the stability of paints and coatings, in the worst case lead-

Table 4.8: pH ranges of mineral fillers

Filler	pH
Barium sulphate	6 to 9.5
Calcium carbonate	9 to 9.5
Cristobalite	8.5
Dolomite	9 to 9.5
Precipitated silicic acid	3.5 to 9
Mica	6.5 to 8.5
Kaolin	3.5 to 11
Kieselguhr	6.5 to 10
Pyrogenous silicic acid, hydrophilic	3.5 to 4.5
Pyrogenous silicic acid, hydrophobic	3.5 to 11
Quartz	6 to 8
Talcum	8.5 to 10.5

ing to coagulation of the aqueous binder. ISO 787 part 9 describes how to determine the pH of powdered fillers. Measurement is performed using a pH meter in a 10 % suspension of the filler at a defined temperature.

Water-based paints and coatings are overwhelmingly anionic, i.e. in the alkaline pH range. That means it is best to use neutral to mildly basic fillers. Anionic systems that incorporate acidic fillers must include a sufficient quantity of pH regulator to prevent destabilisation of the binder. Basic filler materials also have a buffering effect in corrosion protection systems, and are helpful in passivating metallic substrates. That makes it preferable to formulate cationic systems using mildly acidic to neutral fillers. The pH may deviate considerably even with nominally the same mineral, depending on the filler's origin. In the extreme case, a given mineral can be either basic or acidic.

4.1.4.2 Water-soluble matter

Mineral fillers and the impurities they contain also include small quantities of water-soluble matter, which can sometimes be detrimental. The ions contained in water-soluble matter usually tend to accelerate undesired chemical reactions. They encourage corrosion processes at the boundary between the coating and a metallic substrate; water-soluble matter is also liable to impair the stability of emulsions and even lead to coagulation. ISO 787 part 3 describes the drawing-out of water-soluble matter by hot-water extraction. Once the water-soluble matter is determined, further tests like electrical conductivity and resistivity can be performed. A cold-water extraction method also exists; see ISO 787 part 8.

4.1.4.3 Hydrophobicity

The surfaces of mineral fillers exhibit differing hydrophobic behaviour, depending on their composition. Most fillers have hydrophilic (water-affinitive) properties. That means their surfaces are easily wettable with water, making it easy to incorporate them in aqueous systems. Hydrophobic fillers like talcum deflect water, so dispersants and wetting agents are essential to their incorporation. Fillers like silicic acids are often hydrophobised artificially by means of organic coatings. A hydrophobic surface coating can bring advantages in specific applications; surface-coated fillers find applications in water-repellent coatings for building façades, and corrosion protection. Furthermore, hydrophobic fillers are much more compatible with nonpolar systems. It is relatively simple to determine the surface characteristics of a filler. The filler is added to a test beaker filled with water and mixed thoroughly. After a short settling time, quantitative assessments are made of the filler wetting, clouding of the water, and sediment formation.

4.2 Filler analytics

Identifying inorganic materials like fillers and mineral mixtures in paints and coatings is not without its problems, and quantification of mineral compounds containing silicon is especially involved. It entails methods and complex analytical equipment based on the principles of:
- Microscopy
- Spectroscopy
- Chromatography

Spectroscopy is especially effective for identifying fillers, and microscopy facilitates further detailed analysis to differentiate various forms of the same mineral[6]. These surface analysis methods are based on the principle of exciting filler particles with photons, electrons, or ions. The filler for its part emits neutrons, electrons, or photons, or backscatters electrons and ions, which provide the desired information. Chromatography physically-chemically separates mixtures as a consequence of their differing distributions between a stationary and mobile phase[7, 8]. Two popular methods are adsorption and partition chromatography, which as a rule appear in parallel. Chromatographic analysis of fillers delivers data on their organic and ionic components.

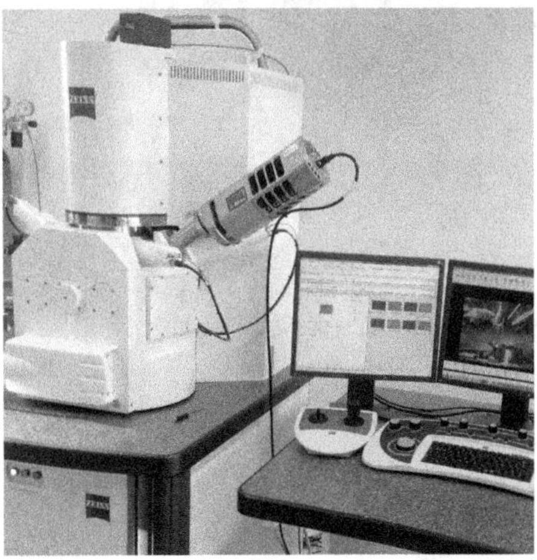

Figure 4.8: Scanning electron microscope (SEM) elemental analysis unit (EDS) source: Omya AG

4.2.1 Scanning electron microscopy

Scanning electron microscopy is an excellent means of imaging filler particles and their surfaces. A variety of methods exist:

- Transmission electron microscopy (TEM)
- Scanning transmission electron microscopy (STEM)
- Scanning electron microscopy (SEM)

Transmission electron microscopy provides a direct image of the sample under investigation. However, because electrons need to transilluminate the sample material, TEM calls for extremely thin samples on the order of nanometres to micrometres (ultra-microtome cuts). The higher the atomic number of the sample, the more difficult it is to obtain images of adequate contrast; alternatively the sample slice has to be made thinner. By detecting electrons that pass through the sample and correlating their timing with points on the sample surface, it is possible to obtain a raster image through the sample. This method is described as scanning transmission electron microscopy. In scanning electron microscopy [9], an electron optics system focuses an electron beam onto the filler surface and sweeps it line-by-line over the sample area of interest. The most common imaging mode detects secondary electrons exiting from the scanned side of the sample to produce a raster image

Figures 4.9 to 4.12: SEM photographs (secondary electron images) of quartz C filler at magnifications of 2,500, 5,000, 7,500 and 10,000-times *source: Omya AG*

of the filler surface at magnifications from 50 to 20,000-times. Backscattered electrons may also be used to detect contrast between areas with different chemical compositions, or form an electron backscatter diffraction (EBSD) image. This is all useful for determining the crystallographic structure of the specimen.

Depending on how the electron microscope is sited, magnifications in excess of 80,000-times are feasible. The prerequisite is a surface with good electrical conductivity; this can be improved by gold sputtering. Depending on how the instrument is equipped, energy dispersive X-ray spectroscopy (EDS) may be used to qualitatively and quantitatively determine elements heavier than beryllium within a defined section of the image. Combining this with other analytical techniques like infrared spectroscopy makes it possible to calculate the composition and proportions of fillers and mineral mixtures. Moreover, the detectable elements and their compounds can be localised by DOT mapping and presented as part of the electron image.

4.2.2 Spectroscopy

Spectroscopy is the study of how electromagnetic radiation (energy) interacts with matter. Spectroscopic methods operate in various wavebands, depending on the energies involved:

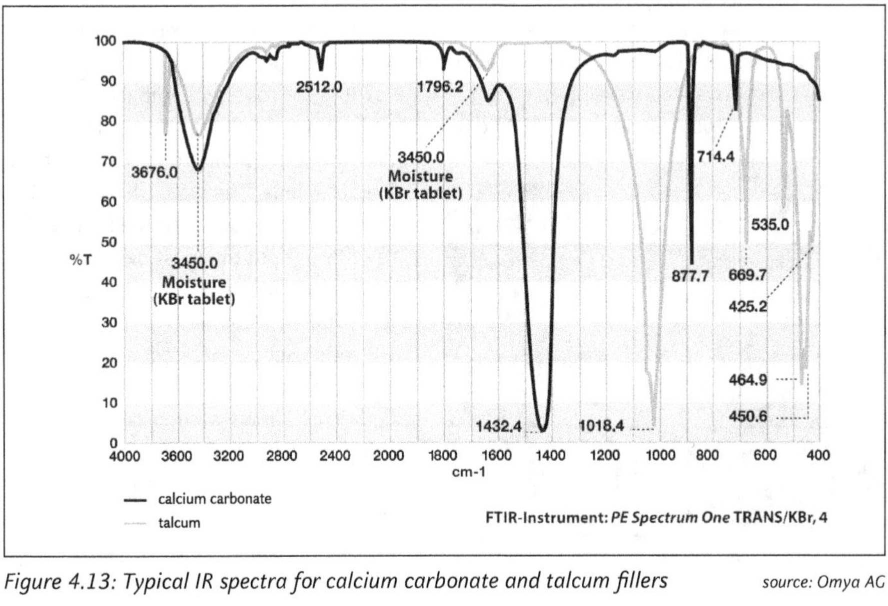

Figure 4.13: Typical IR spectra for calcium carbonate and talcum fillers source: Omya AG

- Infrared spectroscopy (IR/FTIR), optional combined with a microscope
- Atomic absorption spectroscopy (AAS)
- Atomic emission spectroscopy (OES)
- X-ray fluorescence spectroscopy (XRF).

Excepting optical atomic emission spectroscopy (OES) for the moment, spectroscopic analytical instruments are all arranged along similar lines. Light energy at differing wavelengths (IR and element-specific radiation or X-rays) is passed through the prepared sample, or reflected off its surface. The light subsequently reaches an analyser that determines how much light the substance absorbed at a given wavelength. In OES, atoms of the sample material are excited in very high temperature plasma and emit light at their characteristic wavelengths. From the spectrum of the emitted light, it is possible to determine the element concerned. This method is also known as ICP-OES (Inductive Coupled Plasma).

4.2.2.1 Infrared spectroscopy

Infrared spectroscopy was developed by Coblentz as early as 1905 and is in routine use to this day. It is based on molecular excitation by infrared radiation, which causes atomic bonds to vibrate. Depending on molecular structure and atomic movement, valency and deformation vibrations cause the sample under investigation to absorb infrared radiation. As a result, the sample's IR spectra (vibration spectra) contain typical absorption bands; see the examples in Figure 4.13. Single and double beam infrared spectrometers are obsolescent these days, having been superseded by Fourier transform infrared spectrometers (FTIR) that are both faster and provide a higher signal-to-noise ratio [10].

The most advanced development in the field of IR spectroscopy combines FTIR microscopy with a focal plane array (FPA) detector, consisting of an array (typically rectangular) of light-sensing pixels at the focal plane of a lens. The state of the art FTIR technology provides with the highest energy throughput the most detailed chemical information when including attenuated total reflectance (ATR). By means of IR microscopy the most sensitive and fastest chemical imaging solution is applied. This allows a simultaneously collection of thousands of spectra within seconds [11] and offers the scientist a complete solution from single point analysis over mapping to chemical imaging. FPAs, referring to two-dimensional devices that

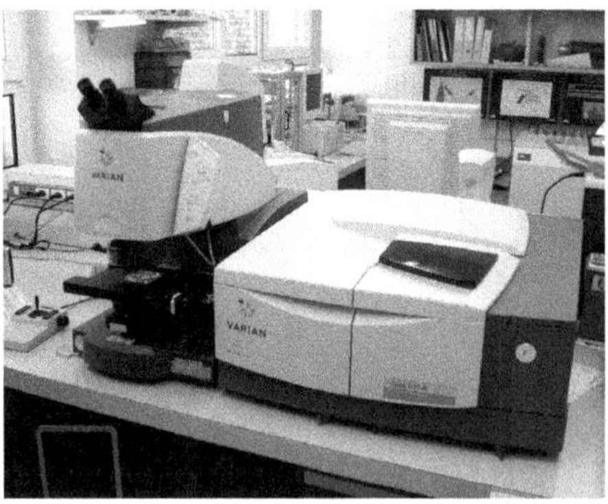

Figure 4.14: FTIR microscopy with focal plane array (FPA) *source: Omya AG*

are sensitive in the infrared spectrum, operate by detecting photons at particular wavelengths and then generating an electrical charge, voltage, or resistance in relation to the number of photons detected at each pixel. This charge, voltage, or resistance is then measured, digitized, and used to construct an image of the material that emitted the photons.

All these instruments help to identify organic and inorganic materials by means of ATR, and deliver qualitative analysis of organic and inorganic fillers and filler blends that have been prepared by pressing into tablet form.

4.2.2.2 Atomic absorption spectroscopy

Atomic absorption spectroscopy (AAS) is a spectral analysis technique for qualitative and quantitative elemental analysis via absorption of optical radiation by free gaseous atoms [12]. AAS is physically based on Bunsen and Kirchhoff's law, which states that an atom is capable of reabsorbing the specific radiation it emits by excitation. This means atoms in the gaseous state will absorb at a specific wavelength and hence diminish the intensity of light passing through the sample. According to the Beer-Lambert law, the measured extinction is proportional to the element's concentration in the analysis sample. AAS instruments operate with a variety of radiation sources (line or continuous-spectrum emitters). Atomisation takes place in a flame, after the sample has been vaporised. Vaporisation and atomisation is done using various equipment, depending on the analysis sample (flame, a graphite tube oven for flameless AAS, electrothermal AAS, hydride AAS and cold vapour AAS).

AAS is used in filler analysis to identify and quantify metals and heavy metals at the trace level. Trace analysis frequently uses the graphite tube technique, which achieves extraordinarily low detection limits. Traces of mercury, arsenic, antimony and tin are analysed by a special flow injection mercury system (FIMS) technique; the cold vapour technique (Figure 4.16) is used to determine mercury traces and the hydride technique is applied to arsenic, antimony and tin. Maximum permissible levels of heavy metals in fillers are defined or limited by their applications, such as painted children's toys, or use in food products. Cleaning

Figure 4.15: Atomic absorption spectrometer (AAS) using a graphite tube, with sample dispenser in the foreground source: Omya AG

agents or human saliva may cause heavy metals present in a coating filler to leach out and be absorbed by humans and the environment. That makes it vital to know exactly what is contained in fillers destined for such applications. There are statutorily regulated conditions (threshold levels) for permitted use.

4.2.2.3 Optical atomic emission spectroscopy

Optical atomic emission spectroscopy (OES) uses various kinds of plasma excitation [13]. Inductive coupled plasma (ICP is one popular excitation source. Here, the fused sample is liquefied, vaporised, and the resulting gaseous atomic cloud is excited to emit element-specific radiation. Atomisation uses an argon plasma torch at temperatures around 10,000 K. The resulting emission spectrum is highly complex.

ICP-OES, see Figure 4.16, is used for simultaneous and sequential multi-element determination of trace metals and non-metals in fillers. This method is deployed for the same reasons mentioned in Chapter 4.2.2.2, atomic absorption spectroscopy (AAS).

4.2.2.4 X-ray fluorescence spectroscopy

The application of fluorescence spectroscopy principles in the X-ray domain is called X-ray fluorescence spectroscopy[14]. The method requires only very small sample quantities for a qualitative and quantitative simultaneous multi-element determination. X-ray fluorescence spectroscopy works by boosting electrons from inner to (higher-energy) outer shells. Other electrons falling back into inner shells fill the resulting vacancies; the difference in energy between the outer and inner shell is emitted as characteristic X-rays. A dis-

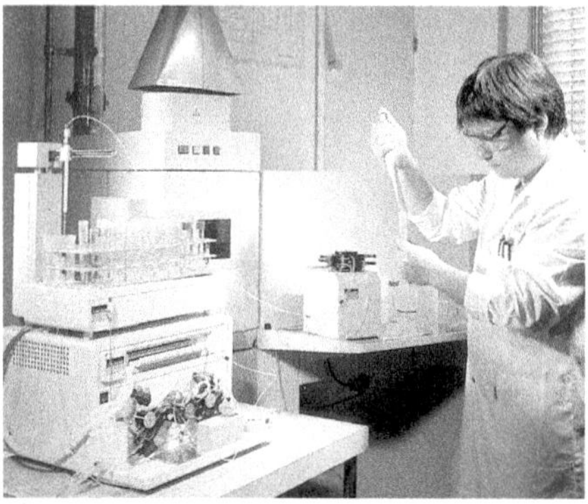

Figure 4.16: Atomic emission spectrometer with inductive coupled plasma (ICP-OES). In the foreground to the left is a flow injection mercury system (FIMS) that uses the cold vapour technique source: Omya AG

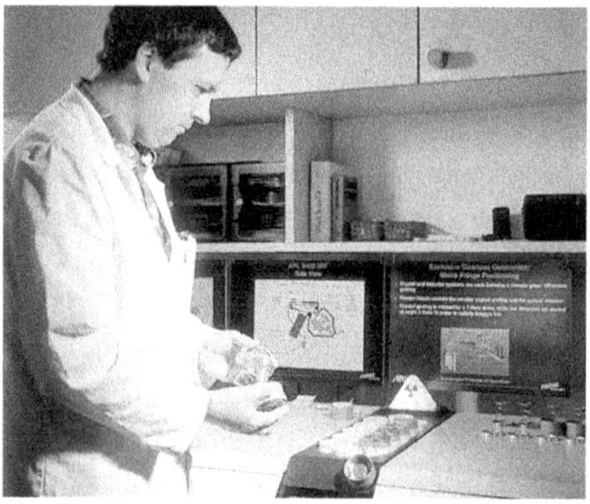

Figure 4.17: Sequential X-ray fluorescence spectroscopy (XRF) for quantitative and semi-quantitative elemental analysis source: Omya AG

tinguishing feature between energy- and wavelength-dispersive XRF: in energy-dispersive X-ray fluorescence, a semiconductor detector measures the energy of the emitted fluorescent radiation. In wavelength-dispersive XRF, an analyser crystal splits the fluorescent radiation into a spectrum of wavelengths.

X-ray fluorescence spectroscopy is suitable for determining elements heavier than beryllium. Ideally, samples can be investigated directly in solid form. Preparation as pressed powder pellets, or embedding in glass discs made from a fusion of lithium tetraborate, is also optimal. This method can provide quantitative determinations of any inorganic filler material.

4.2.3 Chromatography

Chromatography physically-chemically separates mixtures as a consequence of their differing distributions between a stationary and mobile phase [15, 16]. Separation can be accomplished by adsorption or partition chromatography, which as a rule appears in parallel. The following methods are especially relevant to fillers:

- Gas chromatography (GC)
- Ion chromatography (IC)
- Liquid chromatography (LC/HPLC/UHPLC).

Chromatographic analysis cannot identify the filler itself. However, it can analyse the composition of organic material extracted from the filler surface. Such organic material may originate from a hydrophobic coating of fatty acid, milling agents, or dispersants. Fillers and the impurities contained therein may include water-soluble matter. If such matter ionises, it can be determined by ion chromatography.

4.2.3.1 Gas chromatography

Gas chromatography (GC) is a technique for quantitatively separating and identifying vapourable organic compounds [15, 16]. A prepared sample is placed in a separation column contained in a thermostat-controlled oven. A carrier gas transports volatile matter from the sample through the column. Adsorptive and separation chromatography processes split up the various organic components, each of

Figure 4.18: Gas chromatograph with mass spectrometer (GC-MS) for analysing vapourable organic compounds source: Omya AG

which takes a different time to reach the detector at the end of the column. Individual components are identifiable by their retention time in the column, while the amplitude of the detection peak allows quantification of each one.

With a second analytical device block consisting of a mass spectrometer (MS), the GC can be extended to a GC-MS. Due to the different retention time of molecules during elution in the GC, the mass spectrometer breaks each molecule into ionized fragments and detects them using their mass to charge ratio.

Gas chromatography is used to analyse the composition of fatty acids present on fillers. Stearic and other fatty acids are frequently used as surface coatings (hydrophobisation); organic silicon compounds are also a possibility. Other organic milling agents, and occasionally dispersants, are added to minerals during the milling process. Products based on heavier alcohols like glycols, and amine compounds, are widespread.

4.2.3.2 Ion chromatography

Ion chromatography (IC) is the chromatographic separation of (generally inorganic) ion species by their distribution between a stationary and mobile phase [15, 17]. Various separation methods can be used, like ion exchange, ion exclusion and ion pairing. Elution takes place in an electrolyte, whereby individual ions are determined by their respective affinities to the column's stationary phase. Electrical conductivity (measured by a flow control conductometer) is used as a universal detector. An inline sample preparation based on a stopped-flow dialysis does shorten overall analysis time, especially when organic loads are present. The principle of the dialysis is based on the selective diffusion of ions from one liquid (sample or donor solution) to another (acceptor solution) through a membrane. Contrary to dynamic dialysis, where two solutions continuously pass through the dialysis module, the donor solution is temporarily stopped as soon as the concentration in the acceptor solution is the same as that in the donor solution.

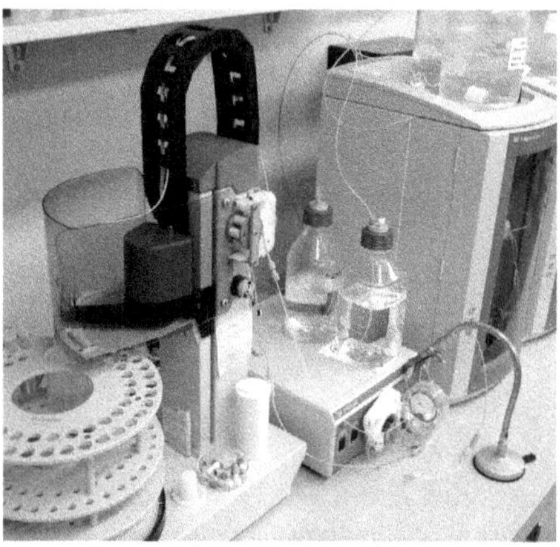

Figure 4.19: Analysis of soluble filler cations and anions by ion chromatography source: Omya AG

Ion chromatography provides data on the type and concentration of soluble filler cations and anions. Fillers should contain low levels of soluble ions wherever possible. High ion concentrations can impair the stability of paints and coatings, especially water-soluble binders and dispersants. Ion chromatography is also used for general quality control of fillers.

4.2.3.3 Liquid chromatography

Liquid chromatography (LC) is a separation technique in which the mobile phase is a liquid, often non-volatile liquids. LC can be carried out either in a column or a plane. Present day liquid chromatography that generally utilizes very small packing particles and a relatively high pressure is referred to as high performance liquid chromatography (HPLC). In the HPLC technique, the sample is forced by high pressure through a column that is packed with irregularly or spherically shaped particles, a porous monolithic layer (stationary phase) or a porous membrane by a liquid (mobile phase) at high pressure in order to identify, quantify and purify the individual components of the sample.

The analysis demand for continuously increasing number of samples led to the latest technology step – ultra high performance liquid chromatography (UHPLC). The speed and efficiency is increased using column materials of less than 2 μm and much higher pressure of almost 1000 bar. This requires reliable ultra high pressure resistant instrument components such as injectors, pumps and valves. UHPLC is often paired with a mass spectrometer.

4.2.4 Further methods

Fillers and minerals are also investigated by methods other than spectroscopic and chromatographic analysis. X-ray diffraction is one method of particular significance to geologists, and vital to the preparatory phase before raw materials are mined to produce fillers. Thermogravimetry is a further important aspect of characterising fillers. Traditional methods like titrations continue to have their place in filler analysis.

Figure 4.20: Ultra High Performance Liquid Chromatography (UHPLC) source: Omya AG

4.2.4.1 X-ray diffractometry

An X-ray diffractometer measures diffraction phenomena in the X-ray part of the spectrum (X-ray diffractometry)[18, 19]. It works by determining the location and intensity of reflections caused by diffraction of X-rays on crystals at defined locations. The reflections can be measured by diffractometer methods:

- Four circle diffractometer
- Imaging plate diffractometer
- CCD diffractometer

The latter two methods allow simultaneous capture of multiple reflections, which brings considerable time savings. X-ray diffractometry (XRD) can be used to investigate single fillers, or blends, because every mineral exhibits a characteristic interference spectrum (peaks). These peaks make a kind of "mineral fingerprint". Peak intensity also allows quantitative determination of minerals present in blends. The intensity depends on chemism, as well as the mineral's structure and concentration. Thus, the raw rock from which fillers are made can be unambiguously characterised for mineral impurities. X-ray diffractometry is therefore primarily a tool for geologists exploring raw material deposits, and only has secondary importance for investigating manufactured fillers. However, when combining an XRD with a deep temperature chamber, the sensitivity to reaction of temperature increases can be studied. This gives certain information about mineral grinding processes.

4.2.4.2 Thermo-gravimetric analysis

Thermo-gravimetric analysis (TGA) means tracing the change in mass of a sample versus temperature or time, under conditions set by a controlled thermal sequence [20]. Known as thermo-gravimetry, this method is applicable only when volatile components escape from the sample during the controlled temperature increase, for example by vaporisation, sublimation, desorption or dissociation, as well as oxidation or reduction. To aid interpretation of thermo-gravimetric curves, many instruments perform first-order time differentiation on the measurement signal and record it, to produce a differentiated thermo-gravimetric curve.

Moisture adsorbed on the filler surface, along with bound crystal water in its crystalline system, will normally escape at temperatures below 450 °C. This is also the zone where any organic compounds present, such as fatty acids, will vaporise. Oxidation can occur at higher temperatures, converting minerals to their oxide forms.

4.2.4.3 Titrations

Titration is a classical wet chemical analysis method [21]. It works by introducing a known quantity of standard solution that induces a specific chemical reaction with an unknown

Figure 4.21: Mineral phase analysis by X-ray diffraction (XRD) source: Omya AG

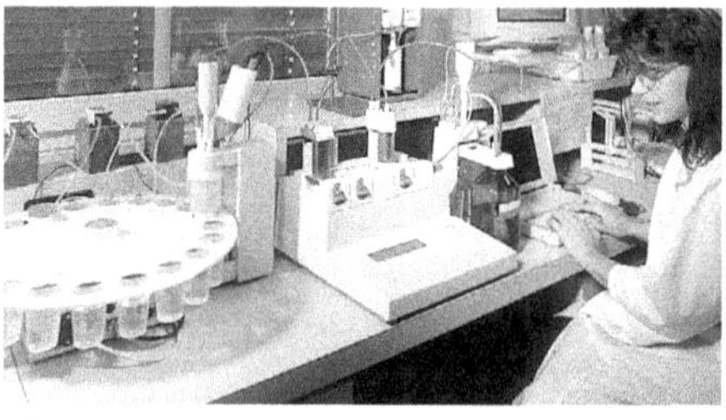

Figure 4.22: Complexometric tests using a Memotitrator source: Omya AG

quantity of a dissolved substance. Determining the titration end-point may be done by chemical indicators, or diverse physical methods: photometry, radiometry, high frequency titration, precipitation analysis and cloud point titration are just a few examples. Electrochemical analysis methods like coulometry are especially important. Titration machines have become commonplace for routine titration work. Complexometric titration and polyelectrolyte titration are mainly used for quantitative analysis of fillers, and any organic accessory compounds they may contain. Use of titration has waned with the emergence of modern and more complex analytical instruments.

4.3 References

[1] ISO Standards Handbook "Paints and Varnishes", Vol. 3 – Raw materials, International Organisation for Standardisation, 2002

[2] Klein, G. A., "Farbenphysik für industrielle Anwendungen", p. 129-149, Springer Verlag, Berlin, 2004

[3] Hering, E, Martin, R., Stohrer, M., "Physik für Ingenieure", 4th edition, p. 409–418, VDI Verlag, Düsseldorf, 1992

[4] Bernhardt, "Particle Size Analysis: Classification and Sedimentation Methods"; Chapman & Hall, London, 1994

[5] Atkins, P. W., "Physikalische Chemie", 3rd edition, Wiley-VCH, Weinheim, 2002

[6] Henzler, M., Göpel, W., "Oberflächenphysik des Festkörpers", 2nd edition, Teubner Verlag, Stuttgart, 1994

[7] Heftmann, E., "Chromatography: Fundamentals and Applications of Chromatography and Related Differential Migration Methods", 5th edition, volume A and B, Elsevier Verlag, Amsterdam 1992

[8] Schwedt, G., "Chromatographische Trennungsmethoden", 3rd edition, Thieme Verlag, Stuttgart, 1994

[9] Goldstein, J. I. et al., "Scanning Electron Microscopy and X-ray Microanalysis", Plenum Press, New York, 1981

[10] Günzler, H., Günzler, H. U., "IR-Spektroskopie", 4th edition, Wiley-VCH, Weinheim, 2003

[11] Varian Documentation, "FT-IR Microscopy and Imaging Solutions", Varian, Santa Clara, 06/2009

[12] Welz, B., Sperling, M., "Atomabsorptionsspektrometrie", 4th edition, Wiley-VCH, Weinheim, 1997

[13] Broekaert, J. A. C., "Analytical Atomic Spectrometry with Flames and Plasmas", Wiley-VCH, Weinheim, 2002

[14] Hahn-Weinheimer, P., Hirner, A., Weber-Diefenbach, K., "Röntgenfluoreszenzanalytische Methoden. Grundlagen und praktische Anwendung in den Geo-, Material- und Umweltwissenschaften", 2nd edition, Springer Verlag, Berlin, 2000

[15] Poole, C. F., "The Essence of Chromatography", Elsevier, Amsterdam, 2003

[16] Kolb, B., "Gaschromatographie in Bildern", 2nd edition, Wiley-VCH, Weinheim, 2003

[17 Weiß, J., "Ionenchromatographie", 3rd edition, Wiley-VCH, Weinheim, 2001

[18] Jenkins, R., Snyder, R. L., "Introduction to X-Ray Powder Diffractometry", Wiley-VCH, Weinheim, 1996

[19] Gluster, J. P., Lewis, M., Rossi, M., "Crystal Structure Analysis for Chemists and Biologists", Wiley-VCH, Weinheim, 1994

[20] Hemminger, W., Cammenga, H.K., "Methoden der Thermischen Analyse", Springer Verlag, Berlin, 1989

[21] Kunze, U. R., Schwedt, G., "Grundlagen der qualitativen und quantitativen Analyse", 5th edition, Wiley-VCH, Weinheim, 2002

5 Properties of fillers

Nature's diversity brings us an abundance of filler materials, a plethora that is further magnified by property variances from one commercially mined mineral deposit to the next. For end users, it all adds up to a huge variety of natural filler materials, one that is amplified further still when one considers the extensive range of chemically equivalent synthetic fillers available, which are usually made by precipitation processes. Olaf Lückert's "Pigment + Füllstoff Tabellen" [1], 192 pages long, gives an idea of the breadth of choice available. While that mammoth work is a compendium of 1952 filler products, this chapter provides a more detailed and comparative review of some typical filler materials in terms of their optical, particle-specific and physical properties. Examples have been chosen to point out the difference in properties that may exist within a given group of filler materials. Chapter 6 will go on to examine the application impact of such differences, whether they occur between members of the same filler group or in the context of mineral fillers in general.

The datasheets that appear throughout this chapter list some generally applicable physical properties for a given filler material, along with a pair of corresponding administrative numbers (CAS and EINECS) that are used to identify and register chemical products (natural fillers included). European readers will note that international CAS numbers (Chemical Abstracts Service) in established use are supplemented by their new EINECS equivalents (European Inventory of Existing Commercial Chemical Substances). CAS numbers are still relevant in non-European applications. Within the European Union, further information about these numbers can be obtained from the government bodies responsible for implementing European law at the member state level, in a process that usually goes into motion shortly after a law has been ratified.

5.1 Carbonates

The choice of carbonate fillers is broad indeed, especially so for the "primary filler", calcium carbonate. The quantity and diversity of commercially mined deposits reflects current demand for carbonates. That in turn stems from carbonates' universal applicability, high brightness, and commercial affordability.

5.1.1 Natural calcium carbonate

Natural calcium carbonate, also known as ground calcium carbonate (GCC), is manufactured in widely differing grades. Products available extend from very coarse particles on the order of millimetres (for architectural plasters and renders), all the way to fine and ultra fine particles in the micrometre range. That makes calcium carbonate eminently suitable for many applications. Whatever its particle size, brightness, or mineral origin, calcium carbonate has certain universally applicable physical properties. The datasheet in Figure 5.1 provides a main summary.

Beyond the points of similarity shown in Figure 5.1, calcium carbonate fillers do differ depending on mineral origin. Raster electron microscopy reveals that chalk, limestone and marble – all of them natural forms of calcium carbonate – are made of differently shaped particles, see Figures 5.2 through 5.4. This is due to their differing geological origins. Even the ground chalk of Figure 5.2 continues to show remains of coccolith shells and other oceanic organisms. The micrograph of limestone in Figure 5.3 can be seen as less dramatic, simply revealing compact particles with fairly rounded, smooth edges. By contrast, the marble in Figure 5.4 exhibits markedly cleaner, sharp fracture edges. They are typical of marble and indicative of its highly crystalline, metamorphic character. Purity requirements for chalk are set out in ISO 3262 part 4, and calcite (limestone and marble) in part 5. The cleanest natural chalk is over 98 % pure, whereas the purest natural calcite is expected to register 99 % or more – a shade grotesque, considering that figure is a whole percentage point higher than the requirement for precipitated calcium carbonate.

Filler	**Natural calcium carbonate (GCC)**
Chemical formula	$CaCO_3$
CAS no.	1317-65-3
EINECS no.	215-279-6
Refractive index	1.59
Mohs hardness	3
Density [g/cm³]	2.7
pH	9
Particle shape	Nodular
Remarks	Soluble in acids, hydrophilic

Figure 5.1: Datasheet for natural calcium carbonate (GCC)

The diversity of calcium carbonate fillers is evident simply from the broad spectrum of fineness grades available. At one end there are popular products with median particle diameters of 2, 5, 10 and 15 µm, see Figure 5.5. Meanwhile, the finest natural grades have median particle diameters under 1 µm. The coarsest products for use in paints and coatings have a top cut up to 100 µm. Coarser products still, with particle sizes ranging up to several millimetres, go into putties, mineral adhesives and sealants, and architectural renders and plasters. Not only the particle size distribution (PSD) itself matters, the gradient of the PSD curve be-

comes an increasing factor with finer products. The steeper the curve, the higher the proportion of particles of consistent, or isodiametric, size. This in turn allows the PSD to approach a particle size down to 0.2 µm, the most efficient dimensional order for optical properties: measurements in this region correspond to half the wavelength of visible light.

Calcium carbonate fillers exhibit distinct variations in their optical properties depending on geographical origin, and most of all, the source mineral. The earliest calcium carbonate industrial fillers were processed from chalk deposits: relatively accessible, and requiring little energy to grind into fine powder. The whiteness of chalk is considerably inferior to that of limestone and marble, which also points to a generally high yellowness index. Yet today's filler applications call for stronger, more neutral whiteness than chalk is capable of. Deposits of white limestone, even marble, are being tapped more and more as a result. Crude limestone, and marble in particular, can be ground to very high and neutral whiteness: see the examples in Table 5.1.

Mineral origin and fineness both correlate to whiteness. The finer the crude rock is ground, the whiter it becomes – at least, down to particle sizes around 1 µm. This having been said, grinding is considerably less influential on whiteness than the material's mineral origin. The high amount of energy involved in further grinding tends toward diminishing returns: the effort expended increases out of all proportion to any improvement in results. Meanwhile, although it is impossible to make the product much finer than it already is, whiteness suffers due to abrasion of the grinding aggregate. Wet grinding is another way to whiten calcium carbonate. This process also accommodates flotation as a means to separate impurities that would otherwise impair whiteness.

Figures 5.2 through 5.4: SEM micrographs (secondary electron images) of three natural calcium carbonate modifications: chalk (5.2), limestone (5.3) and marble (5.4) *source: Omya AG*

Table 5.1: Optical properties of natural calcium carbonate fillers from various sources, with median particle size

Mineral	Origin	Whiteness R_y	Yellowness index	Median particle diameter D50% [µm]
Chalk	France	83.6	9.2	2.2 [1]
Chalk	France	89.2	6.4	1.0 [1]
Chalk	Germany	81.0	9.8	2.8 [1]
Limestone	Spain	91.3	6.4	6.7 [2]
Limestone	Spain	91.9	6.1	4.0 [1]
Limestone	Spain	92.3	5.8	2.6 [1]
Limestone	Thailand	94.0	1.2	2.5 [1]
Marble	Italy	94.4	1.9	11.3 [2]
Marble	Italy	94.9	1.7	4.8 [2]
Marble	Italy	95.1	1.6	2.9 [1]
Marble	China	97.0	0.0	1.6 [1]

[1] Particle diameter determined with Sedigraph 5100 [2] Particle diameter determined with Cilas 920

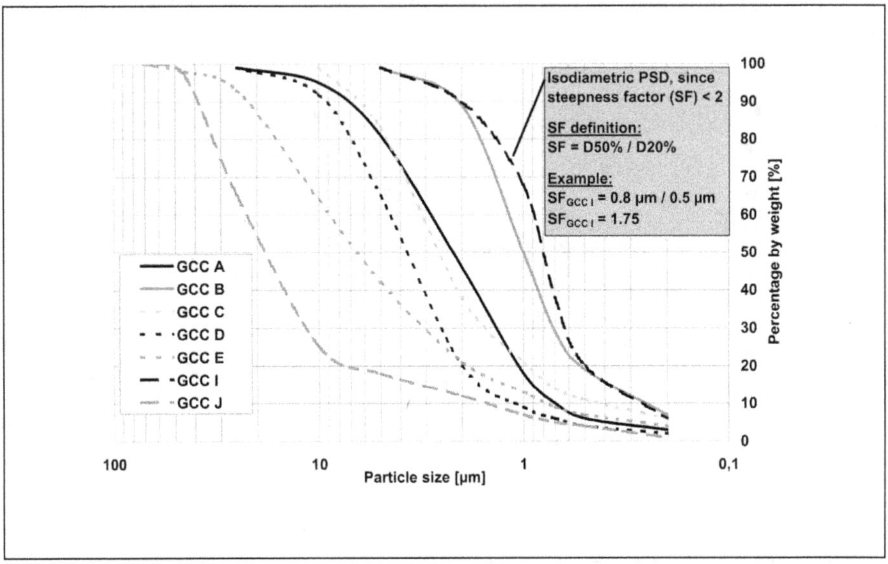

Figure 5.5: Examples of particle size distributions for natural calcium carbonate fillers (GCC)

Figure 5.6 illustrates the impact of filler fineness and mineral origin. Like in Table 5.1, there is a clear whiteness increase and yellowness index decrease as we move from chalk through limestone to marble. Indeed, depending on the source of respective samples, very good quality limestone may in fact be whiter than some marbles, and not so yellow.

Calcium carbonate fillers process excellently in paints and coatings. In aqueous as well as solvent-containing systems, they are easily wettable by standard wetting agents and dispersants. Agglomerations break up fast, with little energy input; dispersal by means of a dissolver is frequently sufficient by itself. Fillers with a steep particle size distribution behave especially well in this regard [2]; stabilising the filler against sedimentation is usually easy as well. Helpful in this respect are calcium carbonate's moderate density of 2.7 g/cm³ and modest binder consumption, even where very fine grades are involved. Expressed in terms of oil absorption, binder consumption is indeed very low compared to other fillers; only barium sulphate comes close. Natural calcium carbonate fillers exhibit oil absorption between 12 and 25 g/100 g of filler. With its compact particles, calcium carbonate exhibits relatively good correlation between oil absorption and specific surface area, see Figure 5.7. Very fine GCC can develop a specific surface area as high as 20 m²/g, although BET specific surface areas between 1 and 10 m²/g are more common. Calcium carbonates possess negligible rheological functionality, this on account of their nodular (rounded) particle structure and low binder consumption. In paints and coatings they act simply as thickeners, encouraging a tighter and less-mobile particle packing density. Calcium carbonate has a very high maximum packing density in water, around 80 %. In contrast to fillers

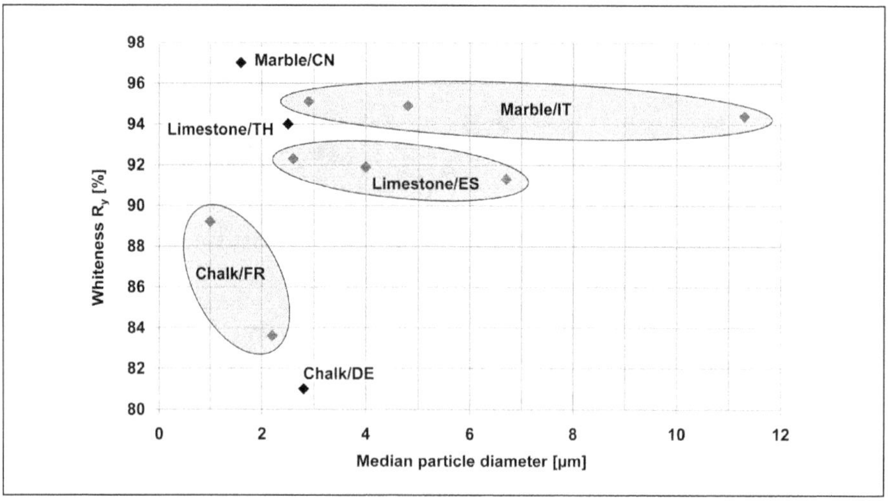

Figure 5.6: Fineness and mineral origin vs. whiteness among natural calcium carbonates

with higher oil absorption, this property permits formulations in which increasing the filler content does not noticeably impair the end product's viscosity, or mechanical properties.

In calcium carbonate applications, it must be appreciated that the material has no acid resistance: at a pH of 6.5 or lower, calcium carbonate will turn into calcium hydroxide and release carbon dioxide. That effectively excludes calcium carbonate from applications that operate in the acidic pH range, for example cathodic electrodeposition (CED) coatings. Yet calcium carbonate is unproblematic in practical use, given that most aqueous paint and coating systems are anionically formulated. Despite being soluble in acids, calcium carbonate finds uses in anti-corrosion applications. Its basic pH acts as a buffer to counteract acids, thus maintaining an alkaline system pH. A basic pH is well known for protecting metallic substrates by passivation, as well as preventing corrosion by hydro-

Table 5.2: Application areas for natural calcium carbonates, along with their typical filler content

Coating system	**$CaCO_3$ content [%]**
Interior emulsion paints, matt	30 to 60
Interior emulsion paints, satin gloss	10 to 20
Exterior emulsion paints	20 to 50
Industrial paints (fillers, primers)	5 to 20
Emulsion-based enamels	0 to 10
Marking paints	30 to 50
Powder coatings	5 to 20
Printing inks	0 to 15
Renders and plasters	60 to 80
Putties	20 to 60

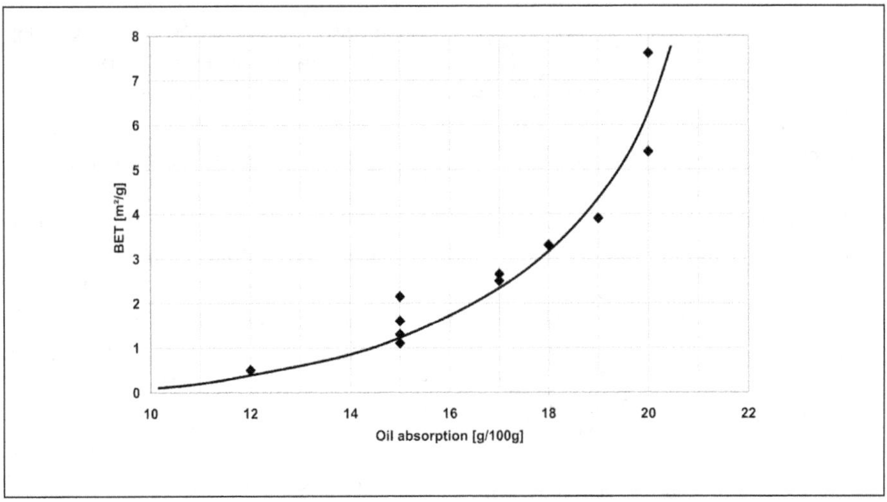

Figure 5.7: Oil absorption of variously fine calcium carbonate, correlated to specific surface area

gen ions[3]. Surface coated calcium carbonate fillers in particular perform well compared to other mineral fillers in Ruf corrosion tests[4, 5]. Calcium carbonate can also seal a surface metallic oxide film by forming a fairly insoluble surface precipitate on the metallic substrate[6]. Coated calcium carbonate is usually treated with stearic or similar fatty acids. Organosilicon compounds are a further possibility, but these have yet to gain much popularity as coatings.

Calcium carbonate presents no health risk, and under the applicable legislation is not classified as a hazardous chemical substance. That means calcium carbonate is innocuous in paints and coatings that come into direct contact with food, and for use in children's toys. Further evidence of calcium carbonate's innocuousness is its approval for use in foodstuffs and pharmaceutical products. These are all properties that suit natural calcium carbonate for use in a great many coating system applications, see Table 5.2.

5.1.2 Precipitated calcium carbonate

Precipitated calcium carbonate (PCC) fillers add to the existing diversity of natural calcium carbonate products on the market. Chemically speaking, precipitated and natural calcium carbonate are indistinguishable. Yet although they share much in common, there are certain, mainly functional divergences between precipitated calcium carbonate and its natural counterpart. The datasheet in Figure 5.8 provides a summary of PCC's general properties.

Precipitated calcium carbonate can have several different particle shapes, depending on conditions at manufacture. Most commercial PCC contains largely calcite (cubic and cigar-shaped) crystalline forms, mixed with a small proportion of aragonite (needle-shaped) crystals. These are visible under the SEM, see Figures 5.9 through 5.11. Scanning electron microscopy also allows precise determination of PCC particle size, even beyond the point where sedimentation analysis and other traditional methods are challenged by particles which are markedly smaller than those of the finest GCC: Brownian molecular motion prevents ultra-fine PCC particles from ever settling completely. This becomes especially problematic with the even finer PCC fillers found in adhesive and sealant applications, (see Figure 5.12).

Filler	**Precipitated calcium carbonate (PCC)**
Chemical formula	$CaCO_3$
CAS no.	471-34-1
EINECS no.	207-439-9
Refractive index	1.59
Mohs hardness	3
Density [g/cm³]	2.7
pH	9.9
Particle shape	Nodular or needle-shaped
Remarks	Soluble in acids, hydrophilic

Figure 5.8: Datasheet for precipitated calcium carbonate (PCC)

Hence, PCC manufacturers frequently measure the Blaine air permeability of a pressed tablet (a technique principally used to determine specific surface area) [7], and derive the median particle diameter from that.

As a rule, paints and coatings use uncoated PCC filler with a median particle diameter between 0.2 and 2 µm. Specified median particle diameters are not always comparable for reasons mentioned above, so it matters a great deal to specify the measurement method used. PCC fillers have a narrower/steeper particle size distribution than GCC grades – in other words, the particles exhibit a strong isodiametric tendency. A tight PSD like this, coupled with median particle size on the order of half the wavelength of visible light, is responsible for PCC's very good optical performance in paints and coatings. Products having high opacity and very neutral whiteness are feasible with precipitated calcium carbonate. Ultra-fine PCC is difficult to manufacture, and this has its economic consequences: PCC fillers can be several times more expensive than GCC.

Figures 5.9 through 5.12: SEM micrographs (secondary electron images) of scalenohedral (5.9) and rhombohedral calcite (5.10) and aragonite (5.11), along with ultra-fine coated rhombohedral calcite (5.12) *source: Omya AG*

Table 5.3: Optical properties, median particle diameter, BET specific surface area and oil absorption of precipitated calcium carbonate (PCC) fillers

Product	Whiteness R_y	Yellowness index	Median particle diameter D50% [µm]	BET specific surface area [m²/g]	Oil absorption [g/100 g]
PCC A	97.5	1.2	0.3 [(2)]	7.2	26
PCC B	97.4	1.1	0.07 [(2)]	14.0	27
PCC C	96.8	2.2	1.1 [(1)]	8.6	41

[(1)] Particle size determined by Sedigraph 5100 [(2)] Particle size determined by air permeability according to NF X11-683

Table 5.3 provides a summary of characteristics for varieties of powdered PCC, showing some impressively high whiteness values and low yellowness index.

Despite its considerably finer particles, PCC's measured BET specific surface area need not by any higher than that of GCC. This is partly attributable to differences in particle shape. Moreover, the coarsest PCC listed in the table above exhibits noticeably higher oil absorption than the two finer PCC types. So PCC exhibits markedly higher binder consumption than GCC, and this should be accounted for in the formulation of paints and coatings: viscosity will be somewhat higher, and mechanical properties may suffer. The latter can be compensated by adjusting the binder content and/or appropriately modifying the filler blend and packing density.

Finer fillers generally enhance dry opacity, or hiding power: see also Chapter 6.4.1. PCC and other, very fine fillers function as spacers between pigment particles and stop them forming agglomerations [8]. The result is a clear boost to optical performance, opacity and colour intensity. Here, it is essential that the filler particles themselves are free of agglomerations, and easily dispersible. In practice, generating the necessary shear forces to disperse PCC agglomerations in solvent-containing systems requires greater energy input than for GCC. Considered another way: under comparable conditions, PCC does not disperse as finely as other mineral fillers [9].

PCC, like GCC, has no significant resistance to acids. Likewise, PCC fillers pose no health hazard; see Chapter 5.1.1 for more details. Finally, Table 5.4 lists a selection of PCC application areas and the respective filler content involved.

5.1.3 Modified calcium carbonate

The latest evolution with calcium carbonate is represented by a filler group called modified calcium carbonate (MCC). In parallel to precipitated calcium carbonate, there are certain differences between MCC and ground calcium carbonate. Besides distinctive functional marks, modified calcium carbonate varies slightly in composition. The datasheet in Figure 5.13 presenting the data for one example of MCC fillers based on roses structures.

Table 5.4: Application areas for precipitated calcium carbonates (PCC), along with their typical filler content

Coating system	**PCC content [%]**
Interior emulsion paints, matt	5 to 20
Interior emulsion paints, satin gloss	4 to 10
Exterior emulsion paints	8 to 12
Industrial paints	4 to 10
Printing inks	0 to 15
Renders and plasters	0 to 10

Modified calcium carbonate owes its uniqueness by its special particle structures. The particle morphology means shape and surface, can be controlled by various reaction process parameter. Versatile particle structures are specified with associated impressions - brain, caviar, rose, golfball and pebble structures are some of them. Others may follow for this new group of functional fillers. The images 5.14 to 5.17 show examples of particles structures available.

The functionality such as absorption characteristics, specific surface area, size of capillaries and pores as well as particle size is also affected by the morphology. For example, the specific surface area BET can range between 10 and more than 100 m²/g, although typical mean particle sizes are clearly above 2 µm. The sophisticated surface structure shows responsibility for the high surface areas. This explains why the degree of freedom in designing inorganic filler particles based on calcium carbonate has been strongly grown. In particular, it is possible to receive an almost uniform particle size with MCC fillers, whereas the standard manufacturing process of GCC always leads to products having a particle size distribution.

Filler	**Modified calcium carbonate (MCC)**
Chemical formula	$CaCO_3$ /$MgCO_3$
CAS no.	1317-65-3 / 546-93-0
EINECS no.	215-279-6 / 208-915-9
Refractive index	1.59
Mohs hardness	3
Density [g/cm³]	2.7
pH	9
Particle shape	Different structures such as brain, caviar, roses, golfballs, pebbles, etc.
Remarks	Soluble in acids, hydrophilic

Figure 5.13: Datasheet for modified calcium carbonate (MCC)

Adding value to paints and coatings is one of the keys for functional modified calcium carbonate fillers. Therefore, the product automatically provides high whiteness R_y combined with a low yellowness index, because the feed material for production is based on marble. The combination of high whiteness and oil absorption enables the formulators, especially of decorative paint systems, to achieve pure white colour tones and high opacity already at medium pvc levels. The opacity function of MCC is smart compared with other opacifiers, which normally tend to reduce mechanical properties such as wet scrub resistance, as their function is duly related to their higher oil absorption. This creates at lower pvc already higher film porosity. This is of course partially similar to MCC fillers, but modified calcium carbonate contributes in addition by its air pores and capillaries. And different to other opacifiers, the amorphous particle structure offers more anchors for the polymer and thus resulting in better wet scrub resistance. This is of course an excellent opportunity to replace partially expensive white pigments, especially in times of high pigment prices or shortages [10, 11].

A particular beneficial property of MCC versus other carbonates is the matting potential. The amorphous particle structure supports very efficient the matting of decorative paints by diffuse light reflection and scattering, while maintaining a high opacity and whiteness at the same time. On one hand this combination is a unique differentiator, because normally opacifier do not contribute positive to matting in paint systems, see also examples of ultra fine GCC fillers. On the other hand, the use of MCC in glossy coating

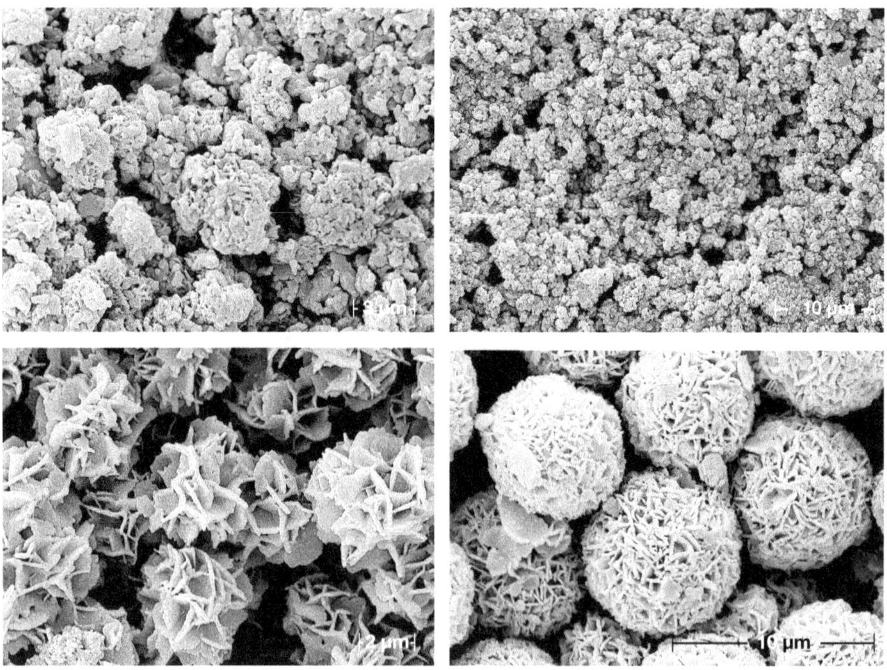

Figures 5.14 through 5.17: SEM micrographs (secondary electron images) of brain (5.14), caviar (5.15), rose (5.16) and golfball structures (5.17) *source: Omya AG*

Table 5.5: Optical properties, median particle diameter, BET specific surface area and oil absorption of modified calcium carbonate (MCC) filler

Product	Whiteness R_y	Yellowness index	Median particle diameter D50% [µm]	BET specific surface area [m²/g]	Oil absorption [g/100 g]
MCC A	94.5	1.7	2.4 [1]	27	50

[1] Particle size determined by Cilas 920

systems is not favourable. In addition, the quite high oil absorption can provide a certain flow control and thickening function, depending on the quantity used in formulations. Depending on the addition level, MCC could possibly replace parts of initial thickeners without destroying the viscosity profile. Adjusting the viscosity at similar initial level also pays off for good mud cracking resistance of the paint films.

Table 5.6: Application areas for modified calcium carbonates (MCC), along with their typical filler content

Coating system	MCC content [%]
Interior emulsion paints, matt	0 to 20
Exterior emulsion paints	0 to 5
Printing inks	0 to 5
Renders and plasters	0 to 5

Despite obviously fragile particle morphology, MCC fillers resist common dispersion methods for decorative paint systems such as high speed dissolver when applying high shear forces. In fact, there is no need for high shear stress, because modified calcium carbonate is easy to disperse without showing agglomerates. Their hydrophilic character in combination with relatively high moisture content prefers water-based applications. Nevertheless, MCC is also used, of course with a lower share, in solvent-based systems. In common with other carbonates, modified calcium carbonates fillers are not resistant to an acidic environment, but also do not present health risks to the environment and workers.

Ideally MCC is used in combination with other mineral fillers, preferably low oil absorbing nodular fillers such as GCC, to optimize packaging. The blending with titanium dioxide is only needed to achieve the minimum required wet opacity in the paint formulation. A very high dry hiding power at high and medium pvc can be adjusted without the addition of white pigments such as titanium dioxide.

5.1.4 Dolomite

Carbonate fillers made from dolomite are not as widespread as those based on natural calcium carbonate. This is attributable to dolomite's comparative infrequency of occurrence;

Table 5.7: Optical figures, median particle diameter, BET specific surface area and oil absorption of dolomite fillers

Origin	Whiteness R_y	Yellowness index	Median particle diameter D50% [μm]	BET specific surface area [m²/g]	Oil absorption [g/100 g]
Norway	93.3	0.6	5.6 [(2)]	2.3	13
Norway	92.4	0.6	3.8 [(2)]	2.5	14
Norway	92.9	0.5	2.1 [(1)]	4.0	16
Poland	94.2	1.3	4.4 [(2)]	2.8	15
Poland	90.6	1.9	2.4 [(1)]	4.1	16
Germany	95.2	0.5	2.3 [(1)]	4.2	17
Germany	94.0	0.8	9.2 [(2)]	1.2	12

[(1)] Particle size determined by Sedigraph 5100 [(2)] Particle size determined by Cilas 920

in property terms it is quite similar to calcium carbonate. Dolomite differs chemically from calcium carbonate due to the presence of a second earth alkali element, magnesium: the rock is composed of calcium and magnesium carbonate in roughly stoichometric proportions. Purity requirements for this double carbonate are defined in ISO 3262 part 7. Especially pure dolomite has a carbonate content in excess of 97 %. The presence of magnesium makes dolomite slightly denser than calcium carbonate – 2.87 g/cm³ – while its Mohs hardness is 3.5.

Dolomite particles are nodular, with a shape comparable to marble modifications of GCC – hardly surprising in the light of the similar metamorphic change undergone by dolomite during its formation. Grinding produces clear fracture edges, indicative of dolomite's crystalline nature. The SEM micrographs in Figures 5.19 and 5.20 show two dolomite fillers from Norway.

Filler type	Dolomite
Chemical formula	$CaMg(CO_3)_2$
CAS no.	16389-88-1
EINECS no.	240-440-2
Refractive index	1.62
Mohs hardness	3.5
Density [g/cm³]	2.87
pH	10
Particle shape	Nodular
Remarks	Acid-soluble, hydrophilic character

Figure 5.18: Datasheet for dolomite

Fillers made from dolomite are known for their neutral white. While calcium carbonates tend to a yellowish white, dolomites are more neutral to bluish, suggesting a plain white shade to the eye. Known deposits of dolomite all over the world have generally superior and quite comparable optical characteristics. R_y whiteness is normally over 90 %. Dolomite's neutral white gives it a yellowness index approaching zero, see Table 5.7.

Dolomite fillers, like calcium carbonate, are supplied in varying fineness grades. The available range, like the material's use, is more restricted due to a relative paucity of deposits and producers. Dolomite is especially popular in the Nordic countries of Europe, where large quantities of the material go into decorative paints. Dolomites exhibit low oil absorption, so their binder consumption is low as well. It is therefore possible to formulate highly filled paints that have satisfactory mechanical properties, while producing only minor impact on viscosity. Dolomite's dispersibility is comparable to that of calcium carbonate and poses no problems for users. Its somewhat higher Mohs hardness has no adverse abrasive effect on machine parts during dispersal.

Table 5.8: Application areas for dolomite, along with their recommended filler content

Coating system	**Dolomite content [%]**
Interior emulsion paints, matt	30 to 60
Interior emulsion paints, satin gloss	10 to 20
Exterior emulsion paints	20 to 50
Industrial paints	5 to 20
Printing inks	0 to 15
Renders and plasters	60 to 80

Dolomite exhibits markedly stronger resistance to cold, dilute acids than does calcium carbonate. All the same, admixture of warm dilute or concentrated acid turns dolomite into the expected acid salts, accompanied by formation of carbon dioxide. In anti-corrosion systems, dolomite can be used in a similar way to calcium carbonate. However, it should be noted that an acid atmosphere containing sulphur dioxide can cause dolomite to form water-soluble salts, accompanied by a change in volume. Like calcium carbonate, dolomite does not present a human health hazard. Nonetheless, high magnesium content precludes its use as an additive in the food and pharmaceuticals industries. Table 5.8 lists some paint and coating applications of dolomite, along with the recommended filler content for each. Economically, dolomite is slightly more expensive to use as a raw material than calcium carbonate. This is compensated on occasion by favourable logistics, as in Scandinavia.

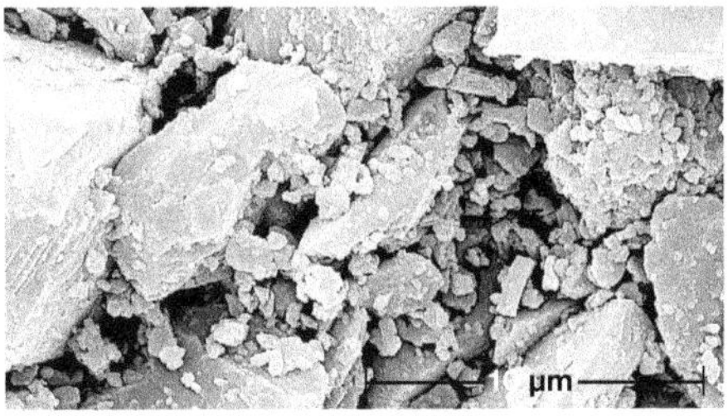

Figures 5.19 and 5.20: SEM micrographs (secondary electron images) of dolomite A and B

source: Omya AG

5.2 Silicates

Silicates occur in an impressive, not to mention confusing multiplicity. Accordingly, users will appreciate some clear definitions. This book narrows the field to focus on the silicates of greatest significance to paints and coatings: talcum, kaolin, mica, feldspar, and precipitated aluminium silicate. Silicates possess functionality that puts them in demand as mineral raw materials, yet they come second in terms of volume consumption. Silicates are indeed "secondary fillers", present in blends to supplement "primary fillers", the pre-eminent one being calcium carbonate. These two filler groups indeed complement each other ideally, not least on account of their respective particle shapes: calcium carbonate is nodular, while silicates are lamellar. Properly combined, these two types of particle produce an even, homogenously packed result.

5.2.1 Talcum

Magnesium silicate hydrate, better known as talcum, has very hydrophobic surface characteristics that make it unusual among the silicates. Adequate supplies of platy talcum are available on the European continent, with processed crude talcum rock from Asia and Australia as backup sources. European deposits do not always match the requirements for coating systems, either on account of their composition or sub-satisfactory optical properties. Chemically speaking, talcum is typically composed of around 61 % SiO_2, 31 % MgO, 5 % H_2O, plus a variety of possible impurities. Requirements for talcum purity/quality are set out in ISO 3262 parts 10 and 11. The datasheet in Figure 5.21 provides a summary of talcum's generally applicable properties.

Filler	**Talcum**
Chemical formula	$Mg_3[(OH)_2/Si_4O_{10}]$
CAS no.	14807-96-6
EINECS no.	238-877-9
Refractive index	1.57
Mohs hardness	1
Density [g/cm^3]	2.75
pH	9
Particle shape	Lamellar
Remarks	Hydrophobic, chemically inert

Figure 5.21: Datasheet for talcum

Talcum's mica-like plate structure indicates its multi-layered arrangement. Talcum is a layer silicate mineral, with strongly hydrophobic boundary edges perpendicular to its layers. The platy particle structure may be more or less pronounced, depending on the deposit concerned – hence the distinction between micro- and macrocrystalline talcum. Talcum is characterised by its aspect ratio, i.e. the relationship of platelet length to thickness. Talcum with an aspect ratio of around 10 : 1 or less is deemed microcrystalline. Aspect ratios considerably higher than this, occasionally as high as 35 : 1, are typical of macrocrystalline talcum. The more platy a specimen is, the more talcum-like its properties. Figures 5.22 through 5.24 show various finenesses of talcum filler with differing aspect ratios. The wafer-thin layer silicate platelets are particularly apparent in Figure 5.24.

Talcum, like carbonate fillers, is available in fine to coarse-particular grades. The finest talcum fillers have a D50% median particle diameter of just less than 1 µm. At such fine particle sizes, though, talcum's lamellar character disappears almost entirely and the particles themselves share closer similarity to nodular fillers. The lamellar structure grows apparent once more with increasing particle size. Determining the size of lamellar talcum and mica particles is not as straightforward as with nodular fillers. The PSD is still obtainable by traditional sedimentation analysis, or by laser diffraction. Figure 5.25 shows examples for commercially available talcum fillers.

Talcum's optical properties are inferior to carbonates in terms of brightness and Ry whiteness, which may dip as low as 60 %. Optical properties also vary considerably between talcum deposits, the result of differing mineral composition and impurities. Selecting high-whiteness talcum fillers has the benefit of more universal applicability. This is especially important in decorative coating system applications, like emulsion paints. In most industrial applications, anti-corrosion systems in particular, whiteness plays if anything a subordinate role.

Talcum's lamellar particle shape gives it a greater specific surface area relative to nodular fillers. This is also reflected in higher oil absorption/binder consumption. Regular ground and classified talcum exhibits good correlation of oil absorption with BET specific surface area.

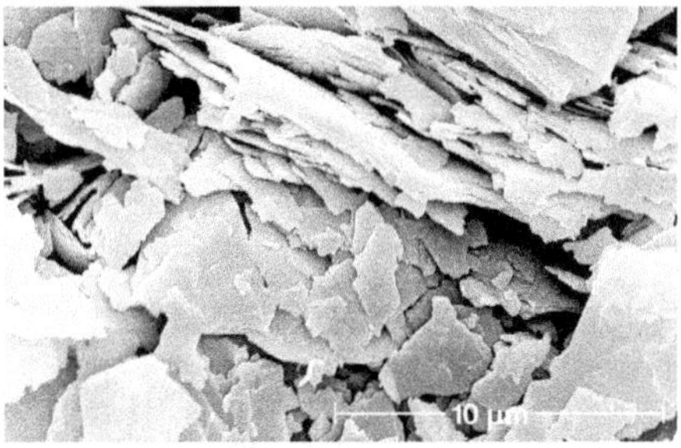

Figures 5.22 through 5.24: SEM micrographs (secondary electron images) of fine talcum with a low aspect ratio (5.22) and coarser talcum fillers with high aspect ratios (5.23 + 5.24)

source: Omya AG

Table 5.9: Optical properties, median particle diameter, BET specific surface area and oil absorption of talcum fillers

Origin	Whiteness R_y	Yellowness index	Median particle diameter D50% [µm]	BET specific surface area [m²/g]	Oil absorption [g/100 g]
Finland	91.4	1.5	2.0 (1)	9.5	42
Finland	86.3	2.3	8.6 (1)	3.1	31
Finland	81.8	1.9	23.4 (2)	1.7	25
France	81.0	2.1	4.3 (1)	4.2	32
France	86.1	1.2	4.3 (1)	4.5	34
Australia	90.6	3.6	8.9 (1)	4.7	24
Norway	78.0	2.7	2.5 (1)	9.7	41

(1) Particle size determined by Sedigraph 5100 (2) Particle size determined by Cilas 920

Remove the fine particles, though, and oil absorption decreases considerably: the Australian talcum included as an example in Figure 5.26 departs from the general correlation curve and hence clearly reveals the phenomenon. With low oil absorption, more talcum can be put into paints and coatings, which in turn helps to reduce the amount of volatile organic compounds (VOCs) used [12].

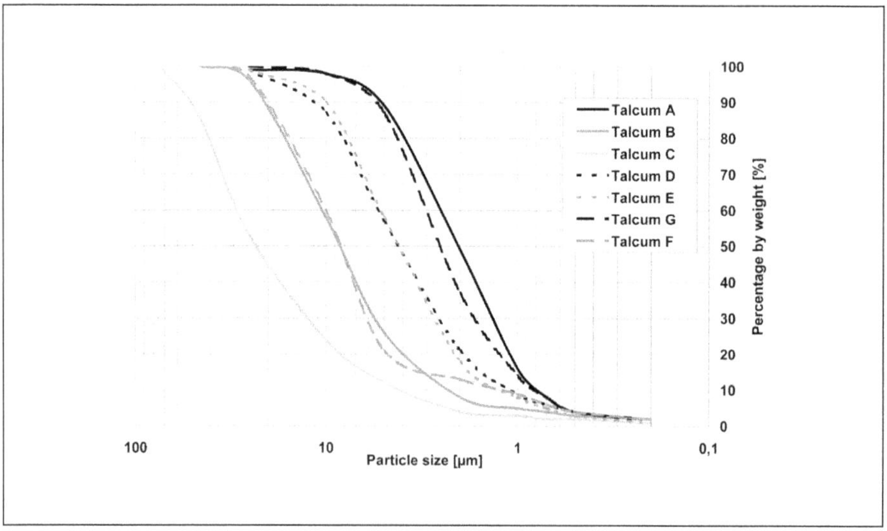

Figure 5.25: Particle size distribution examples for talcum fillers

Talcum's higher binder consumption also effects a reduction in the critical pigment volume concentration (CPVC) in paints and coatings, compared to fillers with lower oil absorption. This translates into greater dry opacity, or hiding power, for a given pigment volume concentration (PVC). In other words, the same dry opacity can be achieved at a lower PVC. As a result, the filler content can be reduced and the binder content increased, which improves wet scrub resistance albeit at higher cost. Talcum fillers, with their hydrophobic, water-repellent surfaces, tend to improve wet scrub resistance [13]. And despite their hydrophobic character, talcum agglomerations can be broken up with the aid of suitable wetting agents and dispersants during incorporation into paints and coatings; a dissolver alone frequently suffices for dispersion. If shear forces are insufficient to produce dispersion, it may be necessary to use a bead mill.

Talcum strongly affects the rheology of paints and coatings on account of its lamellar particle structure. Talcum platelets encourage shear thinning behaviour, so they also play a part in coating characteristics prior to application: storage stability, flow, film formation, etc. Platy particles also pack evenly in combination with nodular fillers. In practical terms, this increases the coating dry film thickness, which in turn enhances application reliability. Talcum is also used as a matt/gloss regulator in paints and coatings: lamellar filler particles scattered at various orientations on the film surface make diffuse light reflectors. Particle size also plays a crucial role here [13–15]; coating films become more matt as particle size increases.

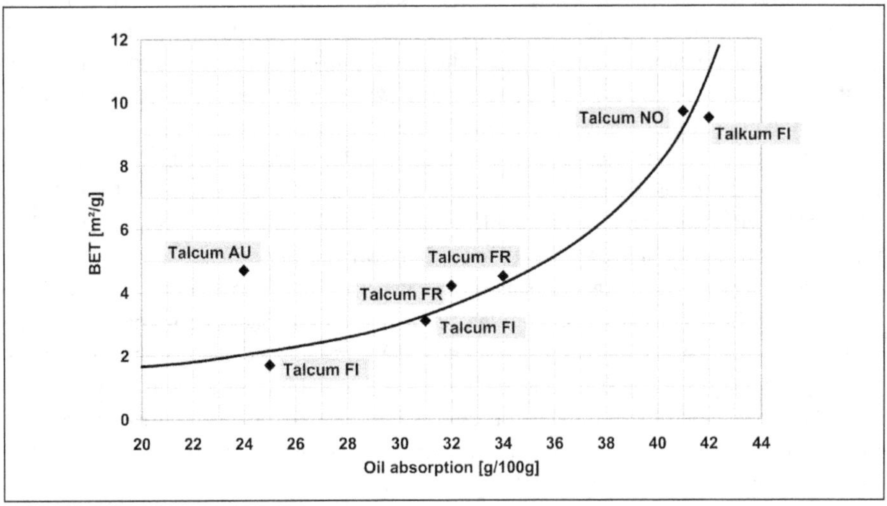

Figure 5.26: Correlation of median particle diameter with BET specific surface area in talcum fillers

Table 5.10: Application areas for talcum, along with their typical filler content

Coating system	Talcum content [%]
Interior emulsion paints, matt	5 to 30
Interior emulsion paints, satin gloss	0 to 10
Exterior emulsion paints	0 to 15
Industrial paints	0 to 20
Anti-corrosion coatings	10 to 30
Emulsion-based enamels	0 to 10
Printing inks	0 to 10
Renders and plasters	0 to 10
Putties	30 to 60

Because talcum fillers behave inertly towards chemicals and acids, they are very popular in anti-corrosion applications. Low porosity is hugely important protection against diffusion, so packing density and the nodular/platy filler blend become especially significant. Fine adjustment of packing density by admixture of appropriately chosen talcum can be crucial to anti-corrosion properties [5]. Talcum also markedly improves adhesion of coating materials to metallic substrates, with corresponding enhancement to their protective properties. With its low Mohs hardness, talcum rubs down very well, making it ideal for paint repairs and sanding between coats in multi-layer systems. Talcum provides extensive functionality demanded by many applications. In coating systems, talcum is very often found blended with nodular fillers to provide optimum benefit.

5.2.2 Kaolin

Kaolin is another member of the layer silicate group of fillers, and is mined at numerous sites around the world. China is probably the best-known producer; it is indeed from here that the alternative name “China clay” originates. The main chemical ingredients of kaolin are SiO_2 at around 46 %, and Al_2O_3 at around 38 %. The quality of natural kaolin fillers is defined in ISO 3262 part 8, which specifies the purest kaolin as containing at least 90 % kaolinite (hydrated aluminium silicate). ISO 3262 part 9 describes calcined kaolins; literature on these thermally treated kaolins occasionally refers to “metakaolin”. Calcined kaolins are expected to contain at least 90 % $Al_2O_3 \cdot SiO_2$, with the highest-quality grades defined as containing no less than 90 % by weight of particles with a fineness of 2 µm. The datasheet in Figure 5.27 lists further generally applicable properties of kaolin.

Kaolins exist in many varieties, not least due to modifications like thermal after-treatment and special coatings. Although kaolin has a less pronounced platy structure than talcum, it still belongs in the group of lamellar fillers. Kaolin exhibits aspect ratios below 15:1 as a rule. Calcination of natural kaolin produces modified particle structures and considerable increases in Mohs hardness. This of course increases abrasion in dispersion equipment and may cause white paints to turn greyer in production. Flash calcination is a special technique that produces closed internal pores. Binders cannot reach these, so opacity tends to increase. According to producers’ specifications, flash-calcined kaolin can exhibit porosity as high as 22 % by volume. Figures 5.28 through 5.30 clearly reveal the differences between natural, calcined and flash-calcined kaolin.

Transmission electron microscopy (TEM) is particularly revealing of calcined and flash-calcined kaolins, with the air pores clearly visible, see Figures 5.31 through 5.33.

Kaolin is available in a fineness range that is comparable to other layer silicates. Commercially available kaolins tend to cluster around median particle diameters between 1 and 5 µm. Although determining kaolin PSD by sedimentation analysis or laser diffraction is not easy, such methods are frequently used in practice. Variations and thermal modifications of kaolin also exist; these have been particularly designed to target optical characteristics like opacity and whiteness. Figure 5.34 shows PSD curves for commercially available natural, calcined and flash-calcined kaolin fillers.

Mineral impurities contained in kaolin deposits mean that their whiteness is nearer 80 % than 90 % as a rule, and accompanied by a high yellowness index in the region of 10 %. Such optical properties are not always satisfactory, especially not as a basis for formulating neutral white paints and coatings. However, calcining brings a marked improvement: the resulting product can sometimes go well above 90 % whiteness, while yellowness index falls to somewhat below 5 %.

At 30 to 40 g/100 g filler, natural kaolins are similar to talcum in their linseed oil absorption. Lower whiteness excepted, the resulting optical performance is not unlike talcum, and may be improved by calcination. Conventional and flash calcination result in markedly increased oil absorption and BET specific surface area, see Figure 5.35. Flash-calcined kaolins are twice as oil-absorbent as their natural counterparts. In paints and coatings, this effects a further reduction in CPVC and a higher binder consumption. The result is relatively high opacity even at low PVC, see also Chapter 6.4.1 [16–18]. Producers claim that this also permits reducing the titanium dioxide content.

Filler	**Kaolin**	**Calcined kaolin**
Chemical formula	$Al_4[(OH)_8/Si_4O_{10}]$	$Al_4[(OH)_8/Si_4O_{10}]$
CAS no.	1332-58-7	92704-41-1
EINECS no.	310-194-1	296-473-8
Refractive index	1.56	1.62
Mohs hardness	1–2	5–6
Density [g/cm³]	2.6	2.6
pH	4.5-7.5	5.5-7.5
Particle shape	Lamellar	Lamellar
Remarks	Chemically inert	Chemically inert

Figure 5.27: Datasheet for natural and calcined kaolin

Binder consumption rises steadily with increasing BET specific surface area and oil absorption. Although this enhances optical properties, it also impairs mechanical properties like wet scrub resistance. To counteract this, the filler's kaolin content can be reduced and the binder content increased accordingly. Although possible in principle, the economic aspects should not be ignored. For this reason, calcined and flash-calcined kaolins are frequently deployed as secondary fillers, at 5 to 10 % of the formulation, to raise opacity ("boosters"). Excessive concentrations of calcined kaolin also cause rapid increases in viscosity. Without corrective measures in the form of suitable thickeners, calcined kaolins are not conducive to high filler content. It should also be noted that most kaolins have an acidic pH. Without careful formulation, this can depress the pH of anionic systems and lead to undesired coagulation of emulsion binders.

Kaolins disperse well with assistance from suitable wetting agents and dispersants. Kaolins are the material of choice in low-PVC emulsion paint formulations, often for satin matt and satin gloss paint films. The traditional European market for kaolin is Great Britain, where emulsion paints are formulated with a considerably lower pigment volume concentration than is the case in continental Europe. Apart from its uses in decorative emulsion paints, kaolin is also an ingredient of industrial paints and enamels. Kaolin's viscosity function helps paints and coatings to brush on in an especially pleasing way. At rest, the filler platelets link up to form a framework; the viscosity that results is especially beneficial to storage stability. The shear forces produced during application reduce this viscosity. A product with well-adjusted rheology will spread well, yet

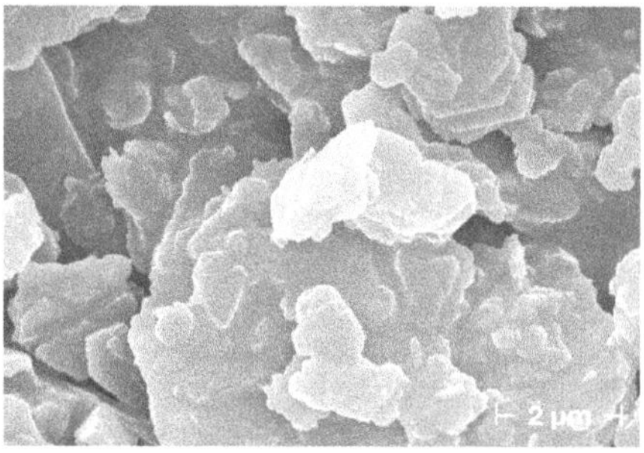

Figures 5.28 through 5.30: SEM micrographs (secondary electron images) of natural (5.28), calcined (5.29) and flash-calcined kaolin (5.30) source: Omya AG

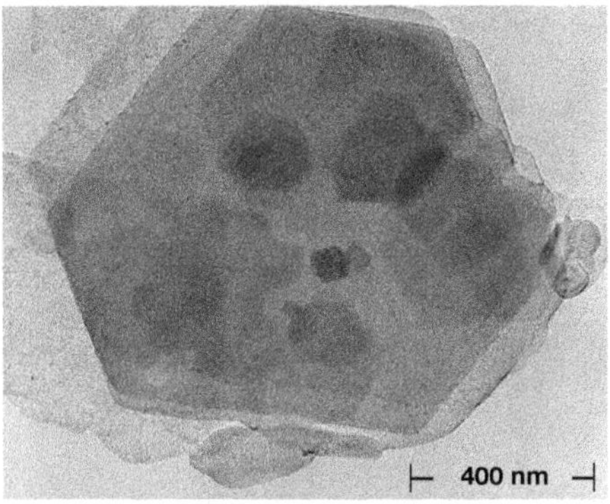

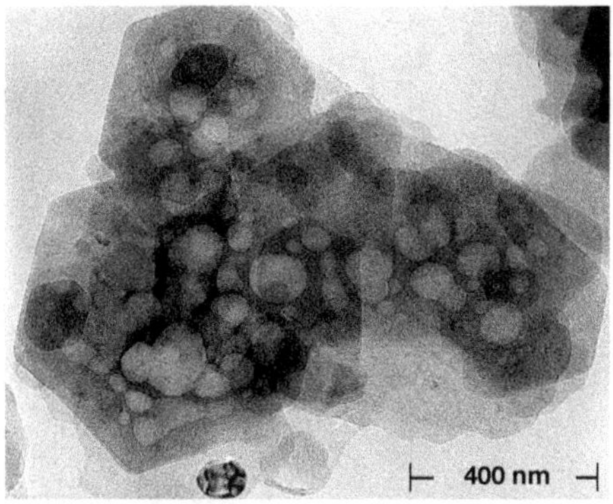

Figures 5.31 through 5.33: TEM micrographs of calcined (5.31) and flash-calcined kaolin (5.32+5.33)

source: Burgess Pigment Company (5.31+5.32), Imerys Minerals Ltd (5.33)

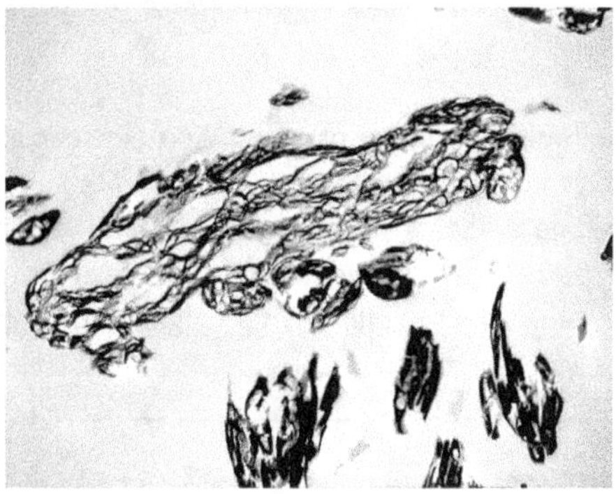

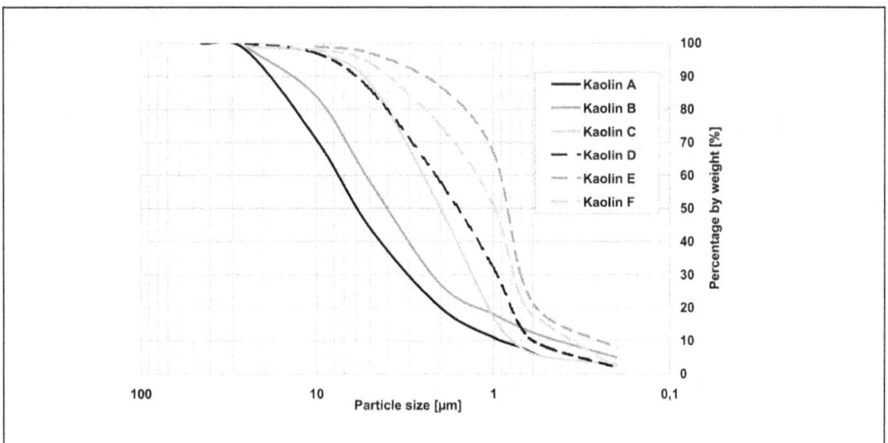

Figure 5.34: Examples of particle size distribution for kaolin fillers

Table 5.11: Optical properties, median particle diameter, BET specific surface area and oil absorption of kaolins

Product	Whiteness R_y	Yellowness index	Median particle diameter D50% [µm]	BET specific surface area [m²/g]	Oil absorption [g/100 g]
Kaolin A	84.7	7.2	5.8[2]	9.4	37
Kaolin B	85.0	10.2	4.2[1]	6.3	31
Kaolin C	94.2	4.8	2.0[2]	6.1	48
Kaolin D	90.6	5.2	1.6[1]	8.6	47
Kaolin E	93.3	4.1	0.8[1]	12.9	78
Kaolin F	91.6	4.1	1.0[1]	11.6	67

[1] Particle size determined by Sedigraph 5100 [2] Particle size determined by Cilas 920

without running or dripping. Another special characteristic of kaolin is its potential for surface coating. The usual fatty acids are suitable, as well as organosilicon compounds like silane that tend to promote adhesion. The functionality of natural and calcined kaolins means they are frequently used as performance-enhancing secondary fillers. This is the reason why the filler content Figures in Table 5.12 are not especially high.

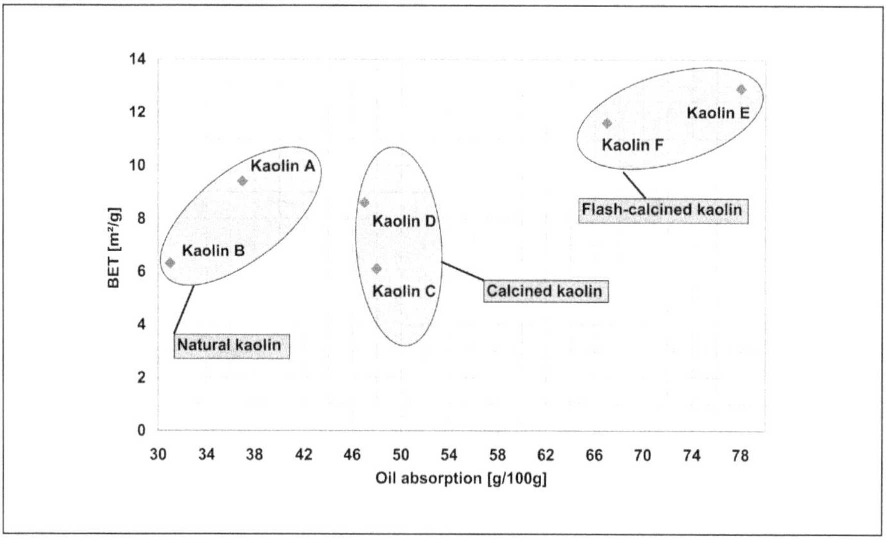

Figure 5.35: Correlation between median particle diameter and BET specific surface area as a function of origin and fineness

5.2.3 Mica

Table 5.12: Application areas for kaolin, along with their typical filler content

Coating system	Kaolin content [%]
Interior emulsion paints, matt	0 to 20
Interior emulsion paints, satin gloss	0 to 20
Exterior emulsion paints	0 to 10
Industrial paints	0 to 10
Emulsion-based enamels	0 to 10
Printing inks	0 to 5
Renders and plasters	0 to 5

Mica-based layer silicates exhibit considerably less variety than talcum and kaolin, and their industrial consumption is lower as well. Nevertheless, users of muscovite mica, the most frequently encountered form, still have a good many product options. ISO 3262 part 12 establishes quality criteria for muscovite mica, with acceptable ranges for its potassium, aluminium and silicon oxide content. Quartz and chlorite may occur as accessory minerals to muscovite in varying concentrations; plastorite is the name given to a mineral consisting of one-third each muscovite, quartz and chlorite. The variety and relative proportions of accessory minerals affect Mohs hardness. This is normally around 2.5 for muscovite mica, with hardness and hence abrasion increasing along with the amount of quartz present. The ISO standard likewise specifies maxima for impurities such as iron and magnesium oxide, all of which can impair whiteness. The datasheet in Figure 5.36 lists generally applicable properties of muscovite mica.

Muscovite mica fillers are available in a wide range of particle sizes. The coarsest have a median particle diameter around 500 µm and the finest just 1 µm. Extra-fine mica fillers are produced by wet grinding. Although this involves more effort than dry grinding, the process is also gentler on the material. That in turn guarantees good delamination and produces extremely high aspect ratios, up to 150:1. Mica has the most pronounced platy structure of all the layer silicates. This favours mica's use in coating systems designed for internal reinforcement, armouring and covering over cracks, while its barrier effect makes for excellent performance in anti-corrosion systems. The conspicuous platyness of muscovite mica and plastorite is apparent in Figures 5.37 and 5.38.

Filler	Mica (muscovite)
Chemical formula	$KAl_2[(OH)_2/AlSi_3O_{10}]$
CAS no.	12001-26-2
EINECS no.	310-127-6
Refractive index	1.56
Mohs hardness	2.5 (Quartz 7)
Density [g/cm³]	2.75
pH	9.5
Particle shape	Lamellar
Remarks	Chemically inert

Figure 5.36: Datasheet for muscovite mica

Table 5.13: Optical properties, median particle diameter, BET specific surface area and oil absorption of mica

Product	**Whiteness R_y**	**Yellowness index**	**Median particle diameter D50% [µm]**	**BET specific surface area [m²/g]**	**Oil absorption [g/100 g]**
Mica A	77.0	3.9	2.9[(1)]	6.3	37
Mica B	74.8	6.1	9.5[(1)]	4.2	21
Mica C	80.7	7.3	5.5[(1)]	5.5	28
Mica D	81.1	5.6	8.0[(1)]	3.8	26

[(1)] Particle size determined by Sedigraph 5100

As with most other fillers, mica's particle size distribution is determined by sedimentation analysis or laser diffraction, although this is not without its problems when platy materials are involved. In emulsion paints, mica fillers of varying fineness can result in very low-matt finishes. In combination with nodular fillers, mica having a properly chosen particle size distribution produces evenly packed particles. This decreases the likelihood of cracking, or alternatively allows application of a thicker coat without the risk of cracks developing. Blending with nodular fillers also enhances wet scrub resistance. Figure 5.39 shows several examples of muscovite mica and plastorite.

Like all naturally occurring layer silicates, mica does not exhibit particularly high whiteness: around 80 % is the norm for muscovite mica as well as plastorite. Corresponding to this at best moderate brightness is the yellowness index, which usually lies between 5 %

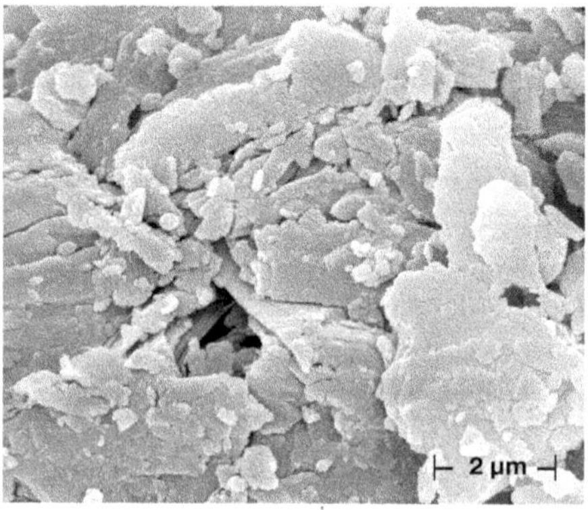

Figures 5.37 and 5.38: SEM micrographs (secondary electron images) of muscovite mica (5.37) and plastorite (5.38)
source: Omya AG

and 10 %. Modest brightness precludes mica as the sole filler in white coating systems. Mica, like the other layer silicates, is thus secondary filler. Yet brightness is not the only factor contributing to this secondary role: in common with other lamellar as opposed to nodular fillers, mica exhibits higher BET specific surface area and oil absorption, although not to the same degree as layer silicates like talcum and kaolin. Table 5.13 shows representative figures for varieties of mica filler.

Table 5.14: Application areas for mica, along with their typical filler content

Coating system	**Mica content [%]**
Interior emulsion paints, matt	0 to 15
Exterior emulsion paints	0 to 15
Industrial paints	0 to 20
Anti-corrosion coatings	10 to 30
Renders and plasters	0 to 10

Mica's platy structure, like that of other layer silicates, also has a rheological function. However, mica produces less shear thinning behaviour than talcum or kaolin. Muscovite mica and plastorite disperse relatively easily with the help of suitable wetting agents and dispersants, so bead mills are not always essential for dispersion. Mica is popular in anti-corrosion systems on account of the barrier effect caused by its pronounced platy structure, as well as its chemical inertness. Moreover, mica tends to improve adhesion, absorbs UV radiation, and is weather-resistant. Thanks to such wide-ranging functionality, mica is found in numerous applications. Table 5.14 lists various application areas, along with the mica content of fillers in use.

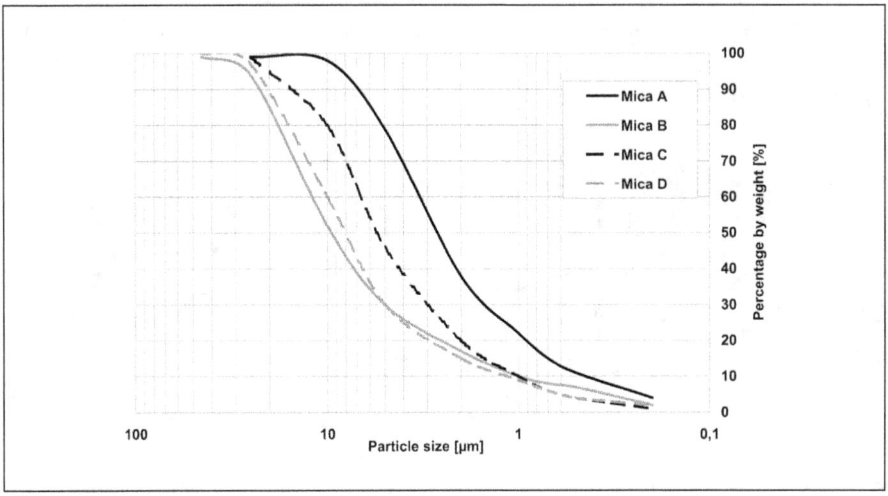

Figure 5.39: Examples of particle size distributions for muscovite mica and plastorite

5.2.4 Feldspar

Feldspar is presented here as the fourth member of the silicate group. In Europe, only relatively small quantities of the material end up in coating systems. Although there are correspondingly fewer commercially available varieties, they are sufficient. In contrast to Europe, feldspar – or to be precise, nephelin syenite – is used more in North American paints and coatings. Nephelin syenite is a special form of the mineral that contains somewhat below 75 % feldspar (albite and orthoclase), and so will be considered with feldspar here. Chemically, feldspar is an alumosilicate comprised of a mineral trio: albite, orthoclase and anorthite. Feldspar is an exception among mineral fillers in that it is not covered by the ISO 3262 standard. Hence, refer to Eurofel, the European Association of Feldspar Producers [19], or Chapter 2.3.4. The datasheet in Figure 5.40 provides a summary of feldspar's generally applicable properties.

Filler	**Feldspar (Albit)**
Chemical formula	$Na[AlSi_3O_8]$
CAS no.	68476-25-5
EINECS no.	270-666-7
Refractive index	1.53
Mohs hardness	6
Density [g/cm³]	2.6
pH	9.5
Particle shape	Nodular
Remarks	Chemically inert

Figure 5.40: Datasheet for feldspar

Feldspar is a tectosilicate, which gives it a structure clearly distinct from layer silicates like talcum, kaolin and mica. Feldspar contains predominantly nodular particles, although analysis always reveals some lamellar particles as well. Figures 5.41 and 5.42 show two feldspars with differing fineness.

Quartz and iron-titanium compounds may occur as accessory minerals in feldspar and affect the grinding process, not least on account of their great hardness. Yet feldspar itself has a Mohs hardness

Figures 5.41 and 5.42: SEM micrographs (secondary electron images) of feldspar with differing fineness
source: Omya AG

Table 5.15: Optical properties, median particle diameter, BET specific surface area and oil absorption of feldspar

Product	Whiteness R_y	Yellowness index	Median particle diameter D50% [µm]	BET specific surface area [m²/g]	Oil absorption [g/100 g]
Feldspar A	90.9	2.1	4.6 [1]	2.5	21
Feldspar B	89.9	2.6	10.8 [1]	1.2	18

[1] Particle size determined by Cilas 920

of 6, making it problematic to grind into very fine fillers. This is the reason behind limited offerings of fine feldspar fillers: the finest products available have a median particle diameter of 2 to 3 µm. Figure 5.43 shows examples of particle size distributions, which can be measured by the usual sedimentation analysis and laser diffraction methods.

In relation to layer silicates, feldspar's whiteness is comparable at just below 90 %. For this reason, feldspar is also used in decorative coating systems such as emulsion paints. Its yellowness index is relatively low as well. Yellowness plays only a subsidiary role in the vast majority of European feldspar applications; its whiteness is adequate for anti-corrosion applications and general building exteriors. Table 5.15 lists typical optical and physical properties for feldspars.

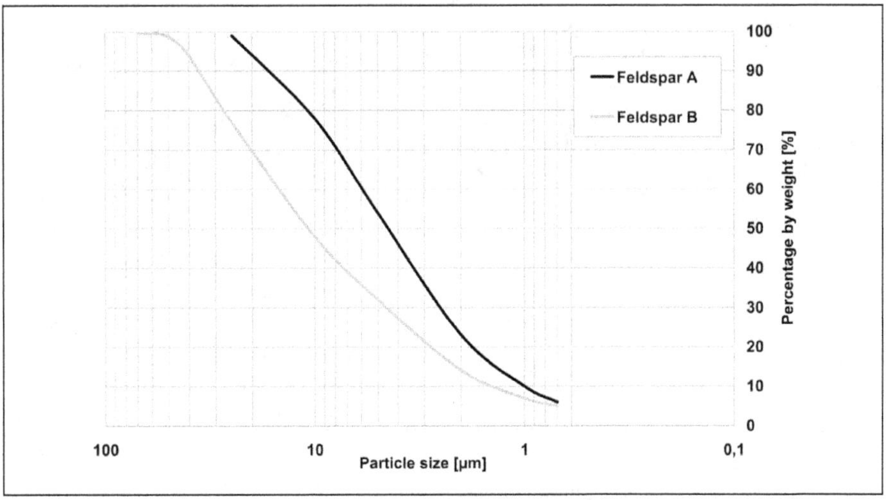

Figure 5.43: Examples of particle size distributions for feldspar

Table 5.16: Application areas for feldspar, along with their typical filler content

Coating system	Feldspar content [%]
Interior emulsion paints, matt	0 to 15
Exterior emulsion paints	0 to 15
Marking paints	20 to 40
Industrial paints	0 to 20
Anti-corrosion coatings	10 to 30

Feldspar exhibits markedly lower BET specific surface area and oil absorption than the layer silicates, due to its predominantly nodular particle shape. This permits formulation of coating systems with higher filler content, thus reducing their volatile organic component. Feldspar's inert nature makes it suitable for heavy-duty anti-corrosion applications. Its relatively high Mohs hardness of 6 is disadvantageous in pale-coloured or white paints and coatings: machine parts tend to suffer abrasion, which can cause greying of otherwise pastel shades. Yet this high degree of hardness also offers advantages in situations that require strong mechanical abrasion resistance, like floor coatings and high-quality emulsion paints. Accordingly, feldspar finds further applications in marking paints that need to be wear-resistant, tough, weatherproof and long lasting. Table 5.16 provides a summary of application areas for feldspar.

5.2.5 Precipitated aluminium silicate

As fillers, precipitated aluminium silicates are in a class of their own. Although technically belonging to the silicate group, they exhibit markedly different properties. As previously mentioned in Chapter 3.3, the full and correct term would be precipitated sodium aluminium silicate. However, precipitated, or synthetic aluminium silicate is the usual industry terminology. Reference to ISO 3262 part 18 should eliminate any possible confusion in this regard; the standard specifies the material's composition and minimum content of amorphous aluminium silicate. The datasheet in Figure 5.44 provides a summary of generally applicable properties.

Filler	Precipitated aluminium silicate
Chemical formula	SiO_2 (with Al_2O_3 und Na_2O)
CAS no.	1344-00-9
EINECS no.	215-684-8
Refractive index	1.46
Density [g/cm³]	2.1
pH	10
Particle shape	Nodular
Remarks	Chemically inert

Figure 5.44: Datasheet for precipitated aluminium silicate

Synthetic production means that commercial precipitated aluminium silicate products are far less diverse than natural fillers. Conventional means of determining particle properties break down with products on this order of fineness; producers of

Table 5.17: Optical properties, median particle diameter, BET specific surface area and oil absorption of precipitated aluminium silicate

Product	**Whiteness R_y**	**Yellowness index**	**Median particle diameter D50% [µm]**	**BET specific surface area [m²/g]**	**Oil absorption [g/100 g]**
Precipitated aluminium silicate A	96.9	0.7	1.2[1]	85	120
Precipitated aluminium silicate B	97.0	0.6	1.0[1]	100	90

[1] Particle size determined by Sedigraph 5100

precipitated aluminium silicate may use the Coulter counter method, for example. Scanning electron microscopy continues to be a highly objective, yet time-consuming approach. All the same, it provides answers about particle homogeneity, which is to say the presence of agglomerations along with their morphology. Precipitated aluminium silicate has a nodular, almost cubic particle shape, see Figures 5.45 and 5.46.

Figure 5.46 shows very homogenously distributed primary particles, with no obvious agglomerations. Figure 5.45 reveals precipitated aluminium silicate composed of highly uniform, small particles measuring approx. 200 nm. The manufacturer specifies this product as having a median particle size of 5 µm according to the Coulter counter method. Measurement by sedimentation analysis indicates a median particle size of 1.2 µm – a discrepancy that serves to emphasise the importance of stating the measurement method

Figures 5.45 and 5.46: SEM micrographs (secondary electron images) of precipitated aluminium silicate at 10,000× magnification, and at 2,500× magnification to check for the presence of agglomerates *source: Omya AG*

Table 5.18: Application areas for precipitated aluminium silicate, along with their typical filler content

Coating system	Precipitated aluminium silicate content [%]
Interior emulsion paints, matt	2 to 5
Interior emulsion paints, satin gloss	2 to 5
Exterior emulsion paints	2 to 5

along with the particle size. From SEM micrographs, it is clear that such product fineness bears strongly on the formulation of paints and coatings. The figures for BET specific surface area and oil absorption in Table 5.17 provide further forceful indication. From this, it is easy to appreciate why the content of precipitated aluminium silicate is limited to 5 %: any more, and viscosity would rise to the point where it can no longer be regulated by reducing the amount of thickener in the formulation.

Precipitated aluminium silicates exhibit very high brightness, with R_y whiteness well over 95 % and neutral yellowness index. This combination of high whiteness and small particle size effects a marked increase in dry opacity and brightness in emulsion paints, achieved through greater film porosity on account of high oil absorption. At the same time, the material maintains optimal spacing between particles of titanium dioxide pigment, and prevents re-agglomeration after the paint is manufactured [20]. This in turn may allow a reduction in the content of expensive titanium dioxide pigments. As well as delivering very good optical performance, the ultra fine particles also function as an anti-settling agent. Precipitated aluminium silicates are easily dispersed and thus aid pigment dispersion as well. Table 5.18 provides a summary of application areas for precipitated aluminium silicates.

5.3 Silicas

Silica fillers are often used for meeting requirements that are completely different from those associated with carbonates and silicates. Characteristic of silica is its chemical inertness and hardness, so it is highly abrasion-resistant. Very fine particles of silica frequently exhibit rheological and optical functionality. They have very specific properties, and correspondingly selective uses.

5.3.1 Quartz

Construction chemical products are large consumers of quartz fillers. While particle size distributions for such applications are rather coarse by paint and coating standards, certain fine grades are available as well. The purity of manufactured fillers is covered in ISO 3262 part 13. The purest is grade A, containing at least 97 % SiO_2. The datasheet in Figure 5.47 lists further generally applicable properties of quartz.

With quartz deposits plentiful everywhere, quartz sand is in very large supply. But quartz is a hard mineral, so grinding, for example, is highly energy-intensive. This may narrow the available offerings, but these nonetheless cater to all needs. The crushed sand used in coating systems contains nodular particles, with sharp fracture edges resembling those of marble. Figures 5.48 and 5.49 provide an impression of the particle morphology of differing quartz grades, as seen by the scanning electron microscope.

Due to their hardness, the finest quartz fillers available have a median particle diameter no smaller than 2 µm. The usual sedimentation analysis and laser diffraction methods are used for determining particle size distributions. Coarser grades are often produced by air-jet sieving, much like other mineral fillers. Figure 5.50 summarises the fine quartz grades available, with their particle size distributions.

Quartz particles are nodular, which accounts for the filler material's low oil absorption and BET specific surface area. The figures here are markedly lower than for silicate fillers, but slightly higher than carbonates. It follows that quartz fillers produce no particular rheological effects in paints and coatings, and can be used up to very high filler content levels. Compared to other nodular fillers like limestone and

Filler	**Quartz**
Chemical formula	SiO_2
CAS no.	14808-60-7
EINECS no.	238-878-4
Refractive index	1.55
Mohs hardness	7
Density [g/cm³]	2.65
pH	7
Particle shape	Nodular
Remarks	Chemically inert

Figure 5.47: Datasheet for quartz fillers

Figures 5.48 and 5.49: SEM micrographs (secondary electron images) of differingly fine quartz at 5000× magnification, quartz A (5.48) with D50% of 1.6 µm and quartz B (5.49) with D50% of 9.1 µm
source: Omya AG

Table 5.19: Optical properties, median particle diameter, BET specific surface area and oil absorption of natural quartz fillers

Product	Whiteness R_y	Yellowness index	Median particle diameter D50% [µm]	BET specific surface area [m²/g]	Oil absorption [g/100 g]
Quartz A	88.8	3.2	1.6 (1)	6.8	25
Quartz B	86.2	6.9	9.1 (2)	1.3	19
Quartz C	87.9	5.0	3.4 (1)	4.1	23

(1) Particle size determined by Sedigraph 5100 (2) Particle size determined by Cilas 920

marble, quartz has inferior brightness and a higher yellowness index. A further disadvantage in many applications is the material's high Mohs hardness of 7. This causes severe abrasion in dispersion machinery, with consequent greying that further impairs the coating material's whiteness. Table 5.19 provides an overview of certain key properties for a variety of quartz fillers.

The properties shown make quartz suitable for fillers in the decorative sector: quartz fillers are weather-resistant, and their low oil absorption makes them useful in highly filled decorative coating systems. The particle shape is advantageous in formulations for render-

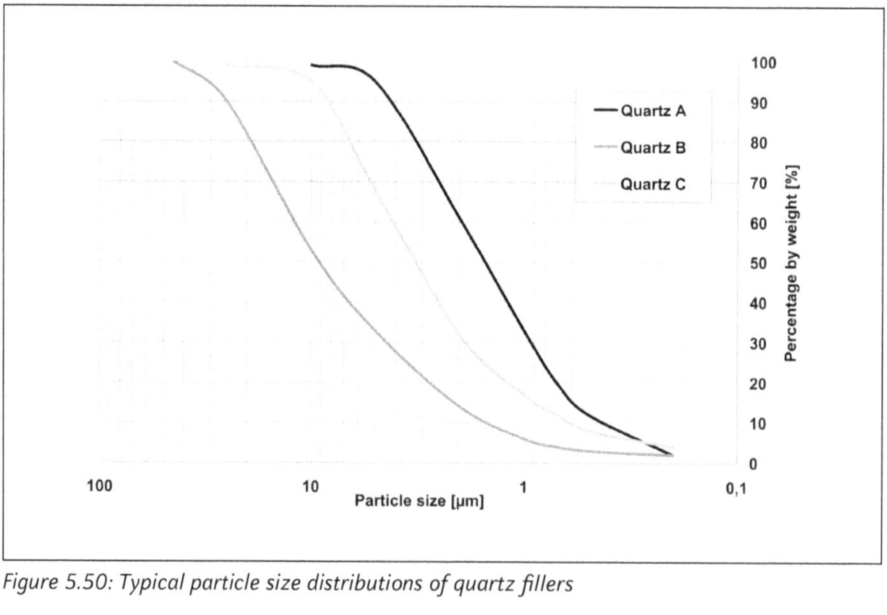

Figure 5.50: Typical particle size distributions of quartz fillers

ing and plastering systems. Coarser particles have a relatively rounded appearance, so they work well in structured and textured renders and plasters. In silicate-based coating systems, silica fillers can also integrate with the binder network by a silification process as a film begins to form. The quartz is thus very strongly anchored, which is ideal for weather-resistant outdoor coatings. Being chemically inert, quartz also finds applications in anti-corrosion systems. Quartz fillers in particular contribute to extremely tough thick film systems, the prime example being epoxy formulations for highly wear-resistant industrial floor coatings. Great hardness is also why quartz is an ingredient of marking paints used in road building and maintenance. Quartz can also enhance reflectivity of road marking paints, for improved visibility at night. Table 5.20 provides an overview of application areas for quartz, with their respective filler content.

Table 5.20: Application areas for quartz, along with their typical filler content

Coating system	**Quartz content [%]**
Interior emulsion paints, matt	0 to 35
Exterior emulsion paints	0 to 20
Marking paints	20 to 40
Industrial paints	0 to 10
Anti-corrosion coatings	10 to 30
Silicate paints	10 to 30

Quartz is non-hazardous to handle if proper protective measures are taken. However, it should be noted that particles of crystalline SiO_2 transported into the lungs (dust particles <7 μm in size) have been shown to cause silicosis, according to the IARC (International Association for Research on Cancer, part of the WHO World Health Organisation) [21]. Moreover, such products have to be labelled as “harmful to health”. The same applies on principle to all mineral fillers that contain more than the statutorily permitted concentration of crystalline SiO_2.

5.3.2 Cristobalite

Cristobalite belongs to the silica group of filler materials. The only kind used as filler in paints and coatings is synthetic cristobalite, made by thermal modification of quartz. Requirements of cristobalite fillers are set out in ISO 3262 part 14, which establishes a minimum cristobalite content of 60 %, with 98 % silica. Cristobalite can only be distinguished from the original quartz by complex x-ray

Filler	**Cristobalite**
Chemical formula	SiO_2
CAS no.	14464-46-1
EINECS no.	238-455-4
Refractive index	1.48
Mohs hardness	6.5
Density [g/cm³]	2.355
pH	8.5
Particle shape	Nodular
Remarks	Chemically inert

Figure 5.51: Datasheet for cristobalite

Table 5.21: Optical properties, median particle diameter, BET specific surface area and oil absorption of cristobalite fillers

Product	Whiteness R_y	Yellowness index	Median particle diameter D50% [µm]	BET specific surface area [m^2/g]	Oil absorption [g/100 g]
Cristobalite A	92.5	0.2	2.6⁽¹⁾	4.0	30
Cristobalite B	95.3	0.9	13.2⁽²⁾	1.2	22

(1) Particle size determined by Sedigraph 5100 (2) Particle size determined by Cilas 920

diffraction analysis. The datasheet in Figure 5.51 shows generally applicable properties of cristobalite.

Cristobalite particles are rounder in shape than quartz, and bear some resemblance to carbonate filler particles. This is also reflected in cristobalite's oil absorption and specific surface area. Nonetheless, both the variety of commercially available cristobalite and the volume consumed are far more restricted than calcium carbonate. Figures 5.52 and 5.53 show particle morphologies for two differing grades of cristobalite filler.

Cristobalite has a Mohs hardness comparable to that of quartz, which imposes similar limitations on the availability of very fine cristobalite fillers. The finest products have a median particle diameter of approx. 2 µm. Cristobalite production is a relatively energy-intensive process, considering that it involves a thermal conversion of quartz and grinding the resultant hard cristobalite. Cristobalite is more expensive than calcium carbonate fillers as a result, which explains why paints and coatings use considerably less of this material. When cristobalite does appear in a formulation, it is only there for very specific

Figures 5.52 and 5.53: SEM micrographs (secondary electron images) of differingly fine cristobalite at 5000× magnification, cristobalite A (5.52) with D50% of 2.6 µm and cristobalite B (5.53) with D50% of 13.2 µm *source: Omya AG*

Table 5.22: Application areas for cristobalite, along with their typical filler content

Coating system	Cristobalite content [%]
Interior emulsion paints, matt	0 to 35
Exterior emulsion paints	0 to 20
Marking paints	20 to 40
Industrial paints	0 to 10
Anti-corrosion coatings	10 to 30
Silicate paints	10 to 30

reasons. Figure 5.54 shows respective particle size distributions for fine and coarse cristobalite filler.

Cristobalite and quartz fillers have quite comparable technical characteristics. They have similar Mohs hardness, along with their very good acid and chemical resistance. Cristobalite fillers are brighter than quartz, and have a very low yellowness index. This opens the possibility of their deployment in light-coloured decorative coating systems, like emulsion paints. Table 5.21 lists some typical key properties. Good weather resistance makes cristobalite suitable for exterior use. It appears as a filler in façade paints, especially in silicate systems. The properties of cristobalite and quartz correlate closely, so refer to Chapter 5.3.1 for further information. Table 5.22 lists application areas for cristobalite.

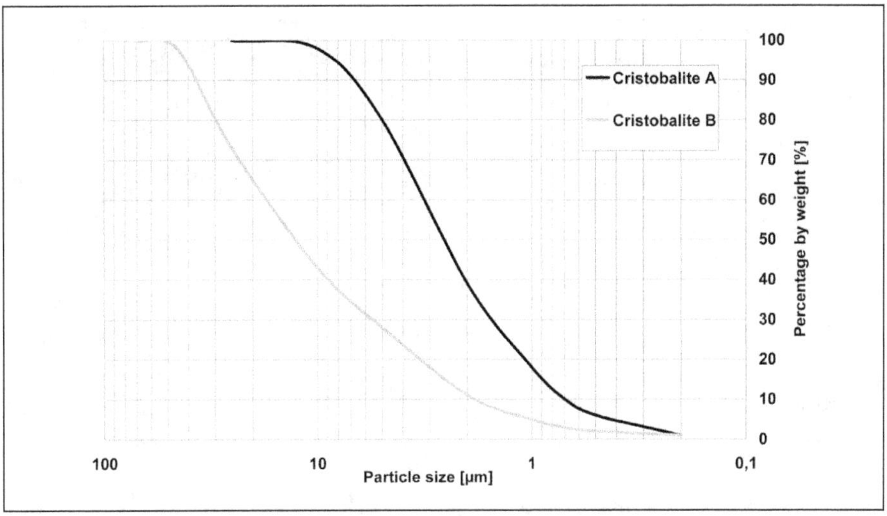

Figure 5.54: Representative particle size distributions for cristobalite fillers

Table 5.23: Optical properties, median particle diameter, BET specific surface area and oil absorption of flux calcined diatomaceous earth fillers

Product	Whiteness R_y	Yellowness index	Median particle diameter D50% [µm]	BET specific surface area [m²/g]	Oil absorption [g/100 g]
Diatomaceous earth A	91.7	2.6	15.5[1]	1.1	67
Diatomaceous earth B	91.1	2.9	6.0[1]	2.0	125

[1] Particle size determined by Cilas 920

5.3.3 Diatomaceous earth

Diatomaceous earth is another filler in the silica group. Natural deposits worth mining are less frequent than those of quartz. Natural diatomaceous earth is often thermally treated (by calcination or flux calcination) to enhance its optical properties, which are no more than average. As a chemical compound, diatomaceous earth has the formula SiO_2. But unlike quartz and cristobalite, it is amorphous and only acquires a crystalline character after calcination. ISO 3262 part 22 sets out the requirements for flux calcined diatomaceous earth. The highest-quality, purest forms of diatomaceous earth contain at least 85 % silica.

Filler	Flux calcined diatomaceous earth
Chemical formula	SiO_2
CAS no.	68855-54-9
EINECS no.	272-489-0
Refractive index	1.45
Mohs hardness	5.5
Density [g/cm³]	2.3
pH	10
Particle shape	Shell shaped
Remarks	Chemically inert

Figure 5.55: Datasheet for flux calcined diatomaceous earth

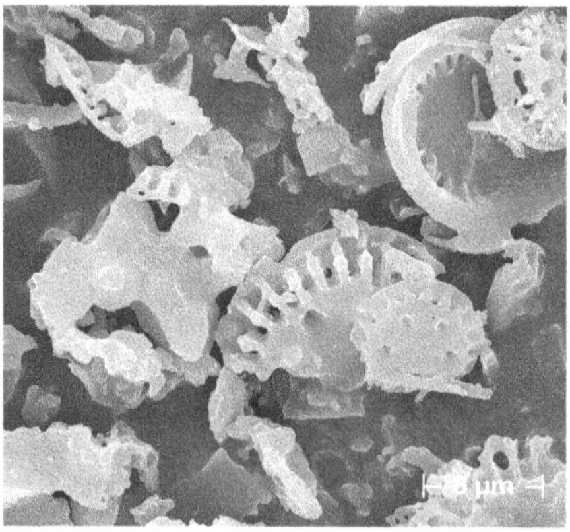

Figure 5.56: SEM micrographs (secondary electron images) of flux calcined diatomaceous earth at 2500× magnification source: Omya AG

The datasheet in Figure 5.55 provides a summary of the material's generally applicable properties.

Natural, calcined and flux calcined diatomaceous earth all have very special particle morphology. Their particles are shaped rather like pillboxes, and filled with capillaries. This is why diatomaceous earth is so porous, despite its relatively modest specific surface area. The large amount of empty space inside the particles is in turn responsible for their high oil absorption. The specific surface area of diatomaceous earth decreases considerably after calcining. The SEM micrographs in Figure 5.56 show the remarkably shaped particles that result.

Table 5.24: Application areas for diatomaceous earth, along with their typical filler content

Coating system	**Diatomaceous earth content [%]**
Interior emulsion paints, matt	0 to 5
Marking paints	0 to 10

Applications for diatomaceous earth concentrate around efficient means of producing matt-finish paints and coatings: it takes relatively little diatomaceous earth to achieve the desired effect. The proportion of diatomaceous earth contained in formulations is no higher than 5 %; more than this causes a steep rise in viscosity due to the material's high oil absorption. Although diatomaceous earth would have a rheological control function, it is not used in this way as a rule. Calcining diatomaceous earth markedly improves its brightness and yellowness index. Table 5.23 shows property profiles of flux calcined diatomaceous earth fillers.

Like silica in general, diatomaceous earth fillers are chemically inert and hence usable in anti-corrosion systems. Diatomaceous earth can also be used in paints, to make them more abrasion-resistant. Table 5.24 shows applications of diatomaceous earth.

5.3.4 Pyrogenous silicic acid

Pyrogenous silicic acids are a distinct sub-class of fillers. Although identical to other silica fillers in terms of chemical structure, their properties differ markedly from natural fillers. Pyrogenous silicic acids are produced synthetically by flame hydrolysis, where amorphous particles are produced during split-second cooling. The process intrinsically produces silica of very high purity, above 99.8 % as defined in ISO 3262 part 20. Figure 5.57 lists generally applicable properties of pyrogenous silicic acid.

The production process turns out ultra-fine, nanoscale spheroid particles. The primary particles range in size from 7 to 40 nm, thereby consisting of some 10,000 SiO_2 molecules each. Yet these particles are not very stable and hence form agglomerations and aggregates, either through interlocking growths or physical-chemical attractive forces. Scanning electron microscopy is hard-pressed as a technique for revealing agglomerations, see Figure 5.58.

Table 5.25: Median primary particle size and BET specific surface area of pyrogenous silicic acids

Product	Median particle diameter D50% [nm]	BET specific surface area [m²/g]
Pyrogenous silicic acid A	16[1]	110
Pyrogenous silicic acid B	7[1]	380
Pyrogenous silicic acid C	40[1]	50

[1] Particle size determined by TEM

Accordingly, producers view pyrogenous silicic acid with the aid of transmission electron microscopy (TEM), a technique that affords excellent resolution at the expense of a strictly two-dimensional view. Figure 5.59 impressively reveals the constituent nanoscale particles of pyrogenous silicic acid.

Given that surface area rises almost exponentially as particle size decreases, it is not hard to imagine the huge specific surface areas of nanoscale primary parti-

Filler	Pyrogenous silicic acid	Pyrogenous silicic acid
Chemical formula	SiO_2	SiO_2
CAS no.	112945-52-5	60842-32-2
EINECS no.	231-545-4	271-893-4
Refractive index	1.46	1.46
Density [g/cm³]	2.2	2.2
pH	4	4
Particle shape	Nodular	Nodular
Remarks	Chemically inert, hydrophilic	Chemically inert, hydrophobic

Figure 5.57: Datasheet for pyrogenous silicic acid

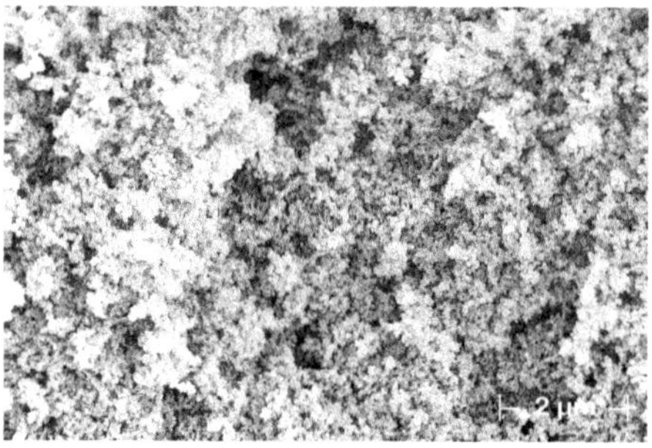

Figure 5.58: SEM micrograph (secondary electron image) of pyrogenous silicic acid A at 10000× magnification source: Omya AG

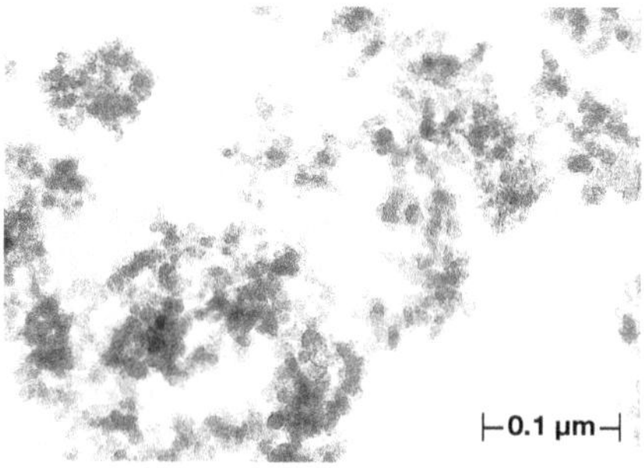

Figure 5.59: TEM micrograph of pyrogenous silicic acid source: Degussa AG

cles and pyrogenous silicic acid agglomerations. The figures for pyrogenous silicic acid are several times higher than for ground mineral fillers. Figure 5.60 clearly illustrates the relationship between median particle size and BET specific surface area [22].

Table 5.26: Application areas for pyrogenous silicic acid, along with their typical filler content

Coating system	**Pyrogenous silicic acid content [%]**
Industrial coatings (HS, coil coating, furniture, effect finishes)	0 to 2
Automotive lacquers	0 to 2
Anti-corrosion coatings	0 to 3
Printing inks	0 to 2
Powder coatings	0 to 1

Pyrogenous silicic acids are dosed into paints and coatings at levels comparable to other additives, so there is no need to further define the optical properties of this white powder. What matters a great deal, though, is the specific surface area. From Figure 5.55 it is apparent that pyrogenous silicic acid must have immense binder consumption to match its huge specific surface area. In fact, it is precisely this property that makes pyrogenous silicic acids into functional fillers for modifying the rheology of coating systems. Table 5.25 lists median particle sizes and specific surface areas for some typical products.

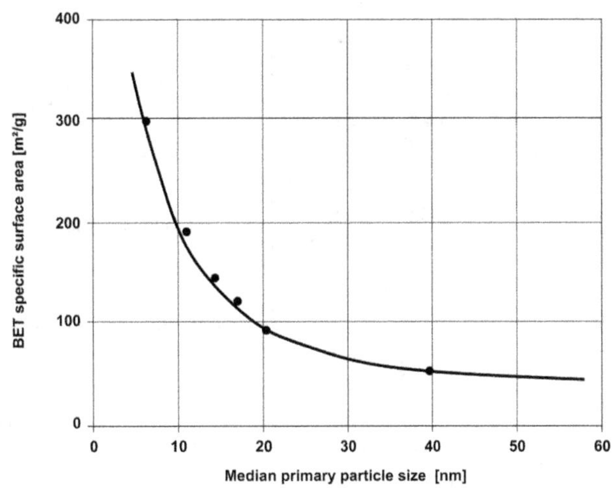

Figure 5.60: Relationship between median primary particle size and BET specific surface area in pyrogenous silicic acid

Pyrogenous silicic acids exhibit very good resistance to acids, but do not stand up well under alkaline conditions. Nonetheless, hydrophobic pyrogenous silicic acids in particular can be effective in anti-corrosion systems[23]. Hydrophobised pyrogenous silicic acids are also relatively easy to disperse, while hydrophilic types require considerably greater shear force. Another factor to consider is these products' long incorporation times, due to their very low bulk and tamped density: the latter is typically in the region of just 0.05 kg/l.

Pyrogenous silicic acid is used to modify rheological behaviour in paints and coatings. Adding the material increases viscosity, and sometimes results in thixotropy. Here, it is suspected that binder molecules bond by strong adsorption to the surface of the pyrogenous silicic acid, which in turn restricts the mobility of polymer chain segments. At rest, the primary particles can also form three-dimensional networks via hydrogen bonds, which subsequently disintegrate through shearing action. Rheological modification by pyrogenous silicic acids may be crucial for properties like sagging resistance, stabilising other solid particles like pigments and fillers against sedimentation in liquid media, boosting pigment colour saturation, and better adhesion and corrosion protection[24]. Hydrophilic types exhibit excellent temperature stability up to 850 °C. Table 5.26 lists application areas with the corresponding dosage of pyrogenous silicic acids.

Unlike fillers made of crystalline silicas like quartz, cristobalite and calcined diatomaceous earths, amorphous particles of pyrogenous silicic acid are not known to cause silicosis or similar health hazards.

Filler	Precipitated silicic acid
Chemical formula	SiO_2
CAS no.	112926-00-8
EINECS no.	231-545-4
Refractive index	1.46
Mohs hardness	6
Density [g/cm³]	2.0
pH	6
Particle shape	Nodular
Remarks	Chemically inert

Figure 5.61: Datasheet for precipitated silicic acid

5.3.5 Precipitated silicic acid

Silicic acids produced synthetically by precipitation represent another sub-class of silica fillers. The silicic acid resulting from this wet process has properties that lie somewhere between pyrogenous silicic acid and natural silica. Precipitation does not attain the same high level of purity found in pyrogenous silicic acid, producing far more impurities such as sulphate, alkali and earth alkali ions. ISO 3262 part 19 details the quality requirements for precipitated silicic acid. Precipitated silicic acids tend to

Table 5.27: Aggregation/agglomeration size and oil absorption for differing types of precipitated silicic acid

Product	Median aggregate/ agglomerate size D50%* [µm]	Oil absorption [g/100 g]
Precipitated silicic acid A	3	220
Precipitated silicic acid B	4	180

* Particle size determined by TEM

contain around 98 % SiO_2, even though the standard calls for a highest purity level of just 95 %. The datasheet in Figure 5.61 lists generally applicable properties.

The precipitated particles have a median primary particle size between 5 and 100 nm, and can be dried by conventional means. The resulting agglomerations may be as large as 100 µm, and these are subsequently ground down to between 1 and 40 µm. A way to obtain fine silicic acid directly is by spray drying after precipitation. Like pyrogenous silicic acid, the precipitated variety has a very high specific surface area on account of its fineness. This is attributable to the outer surfaces in both cases. Yet there are synthetic silica gels, produced by a process similar to precipitated silicic acid, that have even higher specific surface areas. This is the result of primary particles that are smaller still, on the order of 3 to 20 nm, plus their large pores and hence larger inner surface area. Figure 5.62 illustrates the particle morphology of precipitated silicic acid.

Producers of precipitated silicic acids and silica gels use transmission electron microscopy (TEM) to view primary particle morphology, for the same reasons explained in the chapter on pyrogenous silicic acid. SEM technology is used to show aggregations and agglomerations of precipitated silicic acid, see Figure 5.63. Such agglomerations are comprised of very fine primary particles on the order of nanometres, which is why precipi-

Figure 5.62: SEM micrograph (secondary electron image)of precipitated silicic acid A at 10000× magnification source: Omya AG

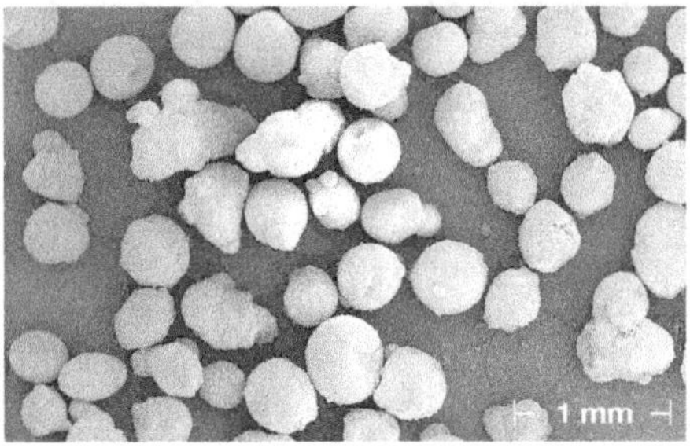

Figure 5.63: SEM micrograph (secondary electron image) of precipitated silicic acid aggregations/agglomerations source: Degussa AG

Table 5.28: Application areas for precipitated silicic acid, along with their typical filler content

Coating system	Precipitated silicic acid content [%]
Industrial coatings (2K, SH, nitro, coil coating, foil)	0 to 15
Enamels	0 to 5
Printing inks	0 to 3

tated silicic acids also exhibit very high binder consumption. Table 5.27 shows the aggregate or agglomeration size determined by SEM along with binder consumption, expressed as oil absorption.

Precipitated silicic acids have similar functional effects as their pyrogenous counterparts. Heavy binder consumption caused by very fine primary particles with a large specific surface area quickly causes the viscosity of paints and coatings to increase. It follows that precipitated silicic acids are used in similar quantities to additives. Precipitated silicic acids are sometimes given an organic coating to improve their suspension behaviour. Coating also improves their dispersibility, as with pyrogenous silicic acids. It takes greater shear forces to break up aggregations and agglomerations in uncoated, hydrophilic precipitated silicic acids. And again, these products can have relatively long incorporation times owing to their very low bulk and tamped density: the latter is typically in the region of just 0.1 kg/l [25].

Precipitated silicic acids are mainly used for mattening coatings across a very broad swathe of applications. By appropriately roughening the surface, they can attain virtually any degree of matt finish. Being chemically inert, precipitated silicic acids are universal in their application. They also make the coating surface less susceptible to mechanical and chemical damage. However, their high binder consumption will increase the viscosity of paints and coatings to a degree that depends on the quantity added / matt finish desired. Precipitated silicic acids and silica gels produced by wet processes usually have comparable applications. Table 5.28 shows application areas and quantities for precipitated silicic acid.

Like pyrogenous silicic acid, the precipitated variety contains no crystalline ingredients. Its amorphous structure precludes known health hazards like silicosis.

Filler	Natural barium sulphate
Chemical formula	$BaSO_4$
CAS no.	13462-86-7
EINECS no.	236-664-5
Refractive index	1.64
Mohs hardness	3
Density [g/cm³]	4.3
pH	8
Particle shape	Nodular
Remarks	Chemically inert

Figure 5.64: Datasheet for natural barium sulphate

5.4 Barium sulphate

Baryte is a natural form of barium sulphate that has long been in use as a filler.

However, barium sulphate has waned in popularity for many applications in recent decades. The falloff was especially drastic around the time when certain paint and coating systems like emulsion paints switched from being sold by weight, to volume. Yet barium sulphate is still widely deployed in many applications where there are no alternative fillers offering comparable technical performance.

5.4.1 Natural barium sulphate

Only high-quality varieties of natural barium sulphate (baryte) are suitable for use in paints and coatings. As with all minerals, the brightness of a given deposit is crucial to the quality of the resulting filler, and hence the range of applications for which it is appropriate. ISO 3262 part 2 sets out requirements for natural barium sulphate fillers and defines the highest-purity grade A material as containing at least 90 % $BaSO_4$. The datasheet in Figure 5.64 lists generally applicable properties of natural barium sulphate.

Natural barium sulphate is produced by grinding pre-crushed crude rock. Grinding results in nodular, rounded particles similar to limestone, with closed surfaces. Figure 5.65 shows a scanning electron microscope view of particle morphology for one type of natural barium sulphate filler.

The range of natural barium sulphate products available is comparable with other natural fillers. The finest ground fillers have a median particle diameter around 1 µm. In coarser products this will typically be 15 µm, or occasionally as high as 30 µm. The particle morphology of natural barium sulphate means that it has a very low specific surface area and hence low oil absorption. The finest products have a BET specific surface area of no more than 4 m^2/g, which results in very low binder consumption of just 20 g linseed oil per 100 g of filler. This allows users to set a correspondingly high filler content, provided the weight resulting from a density of 4.3 g/cm^3 is not an issue. Low oil absorption means that barium sulphate causes hardly any thickening in paints and coatings. However, the high density previously mentioned makes it problematic to stabilise the wet coating and prevent sedimentation of the barium sulphate component. Necessary countermeasures include rheological modifications, improvements to suspension behaviour, and adjustments in the type and quantity of dispersant used. Although natural, white barium sulphate is suitable in principle for emulsion paint formulations, high density puts it at a commercial

Figure 5.65: SEM micrograph (secondary electron image) of natural barium sulphate at 5000× magnification source: Omya AG

Table 5.29: Optical properties, BET specific surface area and oil absorption of natural barium sulphate fillers

Product	Whiteness R_y	Yellowness index	Median particle diameter D50% [µm]	BET specific surface area [m²/g]	Oil absorption [g/100 g]
Barium sulphate A	93.5	1.3	3.2[(1)]	0.6	10
Barium sulphate B	85.0	5.4	6.0[(1)]	0.9	12

[(1)] Particle size determined by Sedigraph 5100

disadvantage. With emulsion paints now sold by volume rather than by weight, using a high-density filler will markedly impair profit margins.

There are distinct variations in brightness between deposits, with natural barium sulphate fillers known to range in whiteness between 60 and 97 %. Very bright fillers are the ideal in terms of universal applicability. Barium sulphate's refractive index of 1.64 is high for a filler, and favourably affects opacity. Although far from achieving properties in the pigment league, the refractive index is nonetheless as much as 0.2 higher than other fillers. Table 5.29 shows comparative properties of two natural barium sulphate fillers.

Barium sulphate is outstandingly inert chemically. It dissolves with difficulty and then to only a minimal extent in concentrated sulphuric acid, and exhibits very good resistance to chemicals in general. These properties stem from a low level of water-soluble salts, and are the principal reason why barium sulphate is popular in marine and automotive anti-corrosion applications. Good lightfastness and weather resistance make barium sulphate suitable for exterior topcoat applications; it is a material of choice for use in exterior powder coatings. Barium sulphate integrates well on account of its high density, and dispersion is likewise unproblematic. Easy dispersibility combined with low oil absorption are reasons for the material's prevalence in highly filled systems like putties and filler pastes. Its low Mohs hardness benefits production of coating materials, since dispersion aggregates do not abrade

Filler	Precipitated barium sulphate
Chemical formula	$BaSO_4$
CAS no.	7727-43-7
EINECS no.	231-784-4
Refractive index	1.64
Mohs hardness	3
Density [g/cm$_3$]	4.4
pH	9
Particle shape	Nodular
Remarks	Chemically inert

Figure 5.66: Datasheet for precipitated barium sulphate

Table 5.30: Application areas for natural barium sulphate, along with their typical filler content

Coating system	**Natural barium sulphate content [%]**
Industrial coatings (anti-corrosion, filler)	0 to 25
Enamels	0 to 10
Marine paints	5 to 25
Putties	20 to 40
Powder coatings	0 to 20

machine parts and it is thus possible to formulate bright colour shades, especially for industrial coatings and enamels. Table 5.30 provides an overview of application areas for natural barium sulphate, with their respective filler content.

Toxicologically, barium sulphate has two distinct aspects. Water-soluble barium salts are considered highly toxic [26], so the amount of water-soluble barium contained in barium sulphate needs to be checked before use. But barium sulphate itself is biologically inactive, and thus deemed non-toxic. It is worth mentioning here that barium sulphate is used to enhance the contrast of medical x-ray images: when ingested it reveals the outlines of internal bodily organs, but is not actually absorbed.

5.4.2 Precipitated barium sulphate

Natural barium sulphate has its synthetic counterpart: precipitated barium sulphate, also known as permanent white (or the French term, blanc fixe). The chemically precipitated product is purer than natural barium sulphate. This is reflected in the requirements of ISO 3262 part 3, which calls for a $BaSO_4$ content of at least 95 %, higher than for the natural product. The datasheet in Figure 5.66 lists generally applicable properties of permanent white.

Particles of precipitated barium sulphate are nodular as a rule, although a platy-shaped form also exists. Permanent white particles have a smooth, closed surface. Precipitated varieties of barium sulphate can be made finer than the natural kind by grinding, which is apparent from the scanning electron micrograph of very fine permanent white shown in Figure 5.67.

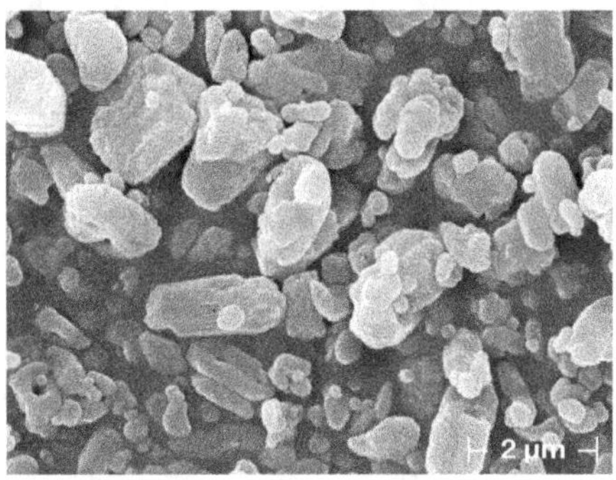

Figure 5.67: SEM micrograph (secondary electron image) of precipitated barium sulphate at 10000× magnification

Table 5.31: Optical properties, BET specific surface area and oil absorption of precipitated barium sulphate fillers

Product	Whiteness R_y	Yellowness index	Median particle diameter D50% [µm]	BET specific surface area [m²/g]	Oil absorption [g/100 g]
Precipitated barium sulphate A	97.1	1.4	0.9[1]	2.7	13
Precipitated barium sulphate B	96.9	1.7	4.0[1]	1.0	12

[1] Particle size determined by Sedigraph 5100

Although available varieties of precipitated barium sulphate are not especially diverse, they are fully adequate for their application purposes. Commercial permanent white products feature very small median particle diameters on the order of 0.7 to 4 µm, along with very low specific surface area and oil absorption. Compared to other fillers, there is no alternative material with a similar degree of binder absorption, which makes precipitated barium sulphate an obvious choice for gloss-finish paints and coatings. It has other benefits too, like a low Mohs hardness of 3 that avoids abrasion of dispersion machinery parts. The chemical process gives precipitated barium sulphate its very high Ry whiteness of around 97 %, which means that permanent white is capable of producing neutral, brilliant white shades. Easy dispersibility in virtually all enamel and emulsion binders is another point in its favour. Along with small-sized particles, this results in a high and lasting gloss. Precipitated barium sulphates are also very stable in wet paints under storage.

Not only does permanent white exhibit good acid and alkali resistance, as a chemically inert filler it is lightfast and weather-resistant as well. That makes it suitable for exterior applications, for example in automotive and industrial coatings, and especially in powder coatings. As well as enhancing optical properties, permanent white also improves the mechanical performance of the paint film. Table 5.31 lists properties of precipitated barium sulphate.

Despite its small particle size, permanent white does not have any major detrimental effect on viscosity. The same applies to use in anti-corrosion applications. Its chemically inert nature, plus very low levels of water-soluble components, are good prerequisites for protecting against corrosion. Only the high density is liable to cause serious sedimentation problems in unbalanced formulations, especially with highly filled systems. Permanent white is frequently blended with other mineral fillers to achieve optimum packing. The resultant coating films have extremely low porosity, blocking diffusion or migration paths for soluble substances that would otherwise encourage corrosion.

As mentioned earlier, precipitated barium sulphate fillers exhibit very low binder consumption. This facilitates formulation of highly filled systems like putties and cavity fillers with closed surfaces for optimal levelling. Emulsion paint applications would be technically feasible as well, but the material's high density precludes them on commercial grounds. Consequently, permanent white is used only in specialised emulsion paints and coatings, such as cationic dispersions in the pH range that cannot be formulated with carbonate fillers, satin-matt emulsion paints, and glossy coatings. Table 5.32 provides an overview of application areas for precipitated barium sulphate.

Table 5.32: Application areas for precipitated barium sulphate, along with their typical filler content

Coating system	**Precipitated barium sulphate content [%]**
Industrial coatings (anti-corrosion, filler)	0 to 25
Enamels	0 to 10
Marine paints	5 to 25
Putties	20 to 40
Powder coatings	0 to 20
Printing inks	0 to 10

Toxicological considerations for precipitated barium sulphate are identical to those for the natural product, see Chapter 5.4.1.

5.5 Aluminium hydroxide and other mineral fillers

There are of course a great many more mineral fillers than those discussed here. Some belong to the major groups – the carbonates, silicates, silica and sulphates – but either they never acquired comparable mainstream prominence, or they are no longer industrially relevant. Grounds for their omission from the present volume include narrow and highly specialised use in paints and coatings, use outside of coating applications that are the focus of this book, peculiarity to the mineralogy of a particular region, or the health hazard they engender. A non-exhaustive list of fillers that will not be considered further includes:

- Wollastonite
- Perlite
- Siliceous earth
- Aerogels
- Silica sole
- Hydrosole
- Calcium aluminium silicate
- Hornblende
- Asbestos
- Calcium sulphate
- Ground slate
- Magnesium hydroxide
- Bentonite
- Sepiolite

The list is far from complete and could easily be expanded. One speciality filler that does, however, merit discussion here is aluminium hydroxide. These fillers are rising in popu-

Table 5.33: Optical properties, BET specific surface area and oil absorption of aluminium hydroxide fillers

Product	Whiteness R_y	Yellowness index	Median particle diameter D50 % [µm]	BET specific surface area [m²/g]	Oil absorption [g/100 g]
Aluminium hydroxide A	95.5	0.9	0.8 [1]	7.1	45
Aluminium hydroxide B	92.4	1.1	10.9 [2]	1.9	26

[1] Particle size determined by Sedigraph 5100 [2] Particle size determined by Cilas 920

larity, in large part because of their flame retardant functionality and applicability in UV-curable coatings.

Aluminium hydroxide fillers are covered by ISO 3262 part 16. The standard relates to synthetic aluminium hydroxide produced by the Bayer process or by direct crystallisation, and natural aluminium hydroxide. As with most mineral fillers, natural aluminium hydroxide is produced by grinding. In practice, the $Al(OH)_3$ content of fillers commonly exceeds 99 % even though the standard does not specify a minimum content or purity for aluminium hydroxides. All it describes is the manufacturing process, and the fineness of the resulting filler. Generally applicable properties are summarised by the datasheet shown in Figure 5.68.

Aluminium hydroxide particles are highly irregular in shape. Whether a specimen contains predominantly nodular or lamellar particles is largely a function of the manufacturing process. The scanning electron micrograph in Figure 5.69 shows the particle morphology of aluminium hydroxide.

Filler	Aluminium hydroxide
Chemical formula	$Al(OH)_3$
CAS no.	21645-51-2
EINECS no.	244-492-7
Refractive index	1.58
Mohs hardness	3
Density [g/cm³]	2.4
pH	9
Particle shape	Nodular to lamellar
Remarks	Chemically inert

Figure 5.68: Datasheet for aluminium hydroxide

With such irregular particles and varying ratios of nodular to lamellar shapes, aluminium hydroxide has a specific surface area and oil absorption similar to silicate fillers. Aluminium hydroxide used in paints and coatings is mostly synthetic: the synthetic variants are both purer and whiter as a result of the precipitation process. Table 5.33 lists properties for varieties of aluminium hydroxide.

Aluminium hydroxide is eminently suitable for producing flame retardant coating materials. Unlike other flame-re-

Table 5.34: Application areas for aluminium hydroxide, along with their typical filler content

Coating system	Aluminium hydroxide content [%]
Emulsion paints	0 to 15
Renders and plasters (organically bound)	0 to 15
UV coatings	0 to 20
Powder coatings	0 to 15

tardants it is halogen-free, and hence an environmentally friendly filler. Aluminium hydroxide fillers are chemically inert and non-toxic. Their flame retardant functionality is due to cooling effects, barrier layer formation, and dilution[27]. The cooling effect is produced by bound water being driven off at around 300 °C. The chemical reaction is endothermic, and as the water is released it vaporises and dissipates further energy. Moreover, thermal decomposition produces no corrosive by-products of the kind that might permanently weaken building structures. The dehydrated aluminium hydroxide proceeds to form a barrier layer that restricts the oxygen supply and formation of gases. Dilution expresses the ratio of aluminium hydroxide to binder, which should be in the region of 150 parts $Al(OH)_3$ to 100 parts binder.

Aluminium hydroxide is a very weak absorber of UV radiation, which makes it ideal as a filler in UV-curable coating systems. Other fillers tend to absorb more strongly, so a coating has to be considerably thinner in order to be properly cured all the way through to the bottommost layer. Table 5.34 provides an overview of application areas for aluminium hydroxide.

5.6 Organic fillers

Organic fillers are a speciality area. The varieties available are produced from a diverse range of chemical compounds: there are plastic hollow spheres made of polyacrylate and puffed-up hollow spheres of vinyl chloride and acrylonitrile, as well as organic polytetrafluoroethylene (PTFE) and polyethylene (PE) fillers. Hollow spheres made of emulsion binders have grown popular in the paints and coatings sector. Figure 5.70 shows a datasheet for one type of polymer hollow sphere organic filler.

Figure 5.69: SEM micrograph (secondary electron image) of aluminium hydroxide at 5000× magnification source: Omya AG

Polymer hollow spheres have two principal functional properties in coating systems: they improve optical performance by increasing their opacity, and reduce their volumetric density. The opacity increase is achieved by the simple and elegant principle of enclosing air, with a refractive index of 1.0, in a sphere of polymer material with a refractive index of around 1.5. The resulting "porosity" stays unfilled by the binder component of the coating system, so there remains a refractive index differential of at least 0.5. When particles of pigment such as titanium dioxide are interspersed by polymer spheres, their effect is to enhance opacity by spacing the pigment particles and avoiding pigment agglomerations. For this to work, though, forces acting on the coating material must not destroy the polymer hollow spheres, which would allow binder to wet the air voids inside. This type of organic filler is particularly effective in paints and coatings that are formulated below the critical pigment volume concentration (CPVC), where the resulting film otherwise contains no trapped air (film porosity) that would contribute to dry opacity. Hence, producers of these (somewhat expensive) hollow spheres argue, it is possible to reduce the content of yet more expensive pigments like titanium dioxide by around

Filler	**Polymer hollow spheres**
Chemical basis	Styrene-acrylate
CAS no.	7732-18-5
EINECS no.	231-791-2
Solid content [%]	51
Median particle size D50% [µm]	0.4
Polymer density as supplied [g/cm³]	1.03
Polymer density as solid [g/cm³]	0.74
Particle shape	Spherical
Pore volume [%]	30

Figure 5.70: Datasheet for a typical polymer hollow sphere organic filler

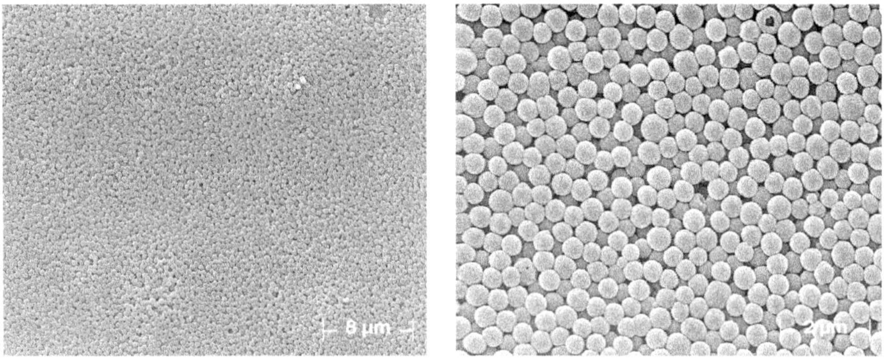

Figures 5.71 and 5.72: SEM micrographs (secondary electron images) of a polymer hollow sphere filler at 2500× and 10000× magnification source: Omya AG

Table 5.35: Application areas for polymer hollow sphere organic fillers, along with their typical filler content

Coating system	**Polymer hollow sphere content [%]**
Interior emulsion paints	0 to 10
Exterior emulsion paints	0 to 10

20 %. Their particle shape and constant, uniform particle size distribution are clearly evident in the scanning electron micrographs of Figures 5.71 and 5.72.

SEM micrographs reveal that the polymer hollow spheres have a very uniform shape; note also the occasional broken sphere as evidence of their "porosity". Yet those broken spheres simultaneously indicate that the polymer material is not immune to all forces. Further evidence of their hollow nature comes from the density differential between the organic filler as supplied, and in the dry state.

With a density that is low to begin with, it is precisely this differential which produces the second effect of using polymer hollow spheres: lowering the density of the coating material itself. This is of particular economic interest in coatings like emulsion paints, which are sold by volume. In practice, adding 8 to 10 % polymer hollow spheres to conventional interior emulsion paint with a pigment volume concentration between 70 and 80 % can reduce the end product's density by around 0.1 kg/l. The commercial sense of using relatively expensive polymer hollow spheres needs calculating on a case-by-case basis and must ultimately be decided by the user. It is a fact that polymer hollow spheres are used more in countries that formulate emulsion paints with a lower PVC, such as Great Britain and the USA. Table 5.35 summarises application areas for organic fillers using polymer hollow spheres.

Today's polymer hollow spheres are eco-friendly, low-odour, and free of formaldehyde and alkylphenol ethoxylates (APEO). The level of residual monomers is very low, and production processes do not use ammonia for neutralisation.

5.7 References

[1] Lückert, O., "Pigment + Füllstoff Tabellen", 6th edition, Vincentz Verlag, Hannover, 2002

[2] Werner, R., "Steep Particle Size Distribution Curves as a Determining Factor in the Use of Fine Extenders in Various Coating Systems", presentation at the annual convention of the Federation of Societies for Coatings Technology (FSCT), New Orleans, 1998

[3] Ruf, J., "Organischer Metallschutz", Vincentz Verlag, Hannover, 1993

[4] Ruf, J., Farbe + Lack, vol. 101, p. 922 ff., Vincentz Verlag, Hannover, 11/1995

[5] Gysau, D., "Influence of Extenders and Extender-Combinations on the Anti-Corrosion Properties of a Water-Thinnable Primer", XXIV. Fatipec Congress, Interlaken, 1996

[6] Strauch, D., Farbe + Lack, vol. 74, p. 1192 ff., Vincentz Verlag, Hannover, 12/1968

[7] Socal documentation, "Precipitated Calcium Carbonate – Production, Properties, Applications", Solvay, Rheinberg, 2001

[8] Dietz, F. P., "Spacing for better effects", p. 14 – 20, European Coatings Journal, Vincentz Verlag, Hannover, 7-8/2003

[9] Gysau, D., Merz, M., Chetelat, E., "OMYACOAT 250- und 850-OG – zwei neue Füllstoffe für Farben und Lacke", p. 18-24, Welt der Farben, Cologne, 4/2003

[10] Gysau, D., "The time is ripe for revolution – Modified Calcium Carbonate (MCC)", XXIX. Fatipec Congress, Ghent, 2008

[11] Gysau, D., "Novel Modified Calcium Carbonates for Decorative Coatings", Middle East Coatings Conference, Dubai, 2010

[12] Penne, R. A. H., "Microcrystalline talcs", European Coatings Journal, Vincentz Verlag, Hannover, 3/1998

[13] Mondo Minerals Technical Bulletin no. 511, "Benefits of Talc Grades of Mondo Minerals in Decorative Paints", Mondo Minerals, Kajaani, 4/2002

[14] Katholnig, P., "Neue umweltverträgliche Wege der Mattierung", VILF publication series, no. 5, 2003

[15] Pauly, J. J., Menez, M., "The use of talc in optimizing coatings formulations", p. 151 – 152, The Coatings Agenda Europe 1997

[16] McGuffog, R., "Lacke optisch verbessern mit Kaolinen", vol. 105, p. 110 – 113, Farbe + Lack, Vincentz Verlag, Hannover, 6/1999

[17] Ashek, L., "Kaolin – neue Füllstoff-generation mit Funktion", vol. 107, p. 46 – 48, Farbe + Lack, Vincentz Verlag, Hannover, 2/2001

[18] Burgess documentation, "Kaolin Clays – Processing & End Uses", Burgess Pigment Company, Sandersville, 6/2004

[19] Eurofel – The European Association of Feldspar Producers, Internet www.ima-eu.org/en/felindex.html, 2004

[20] Degussa documentation, "Sipernat 820 A for Emulsion and Decorative Paints", Technical Bulletin Pigments no. 34, Degussa, Frankfurt

[21] IARC – International Association for Research on Cancer, Internet www.iarc.fr

[22] Degussa documentation, "Basics of Aerosil", Fine Particles publication series no. 11, Degussa, Frankfurt

[23] Kittel, H., "Lehrbuch der Lacke und Beschichtungen", vol. 2, Hirzel Verlag, Stuttgart, 1998

[24] Degussa documentation, "Hydrophobic Aerosil-Types and their Use in the Coatings Industry", Pigment publication series no. 18, Degussa, Frankfurt

[25] Degussa Documentation, "Acematt - Matting Agent for the Coatings Industry",Fine Particles publication series no. 21, Degussa, Frankfurt

[26] Wypich, G., "Handbook of Fillers" p. 40, ChemTec Publishing, Toronto 1999

[27] Wypich, G., "Handbook of Fillers" p. 22 – 24, ChemTec Publishing, Toronto 1999

6 Applications of fillers

This chapter is intended for people who use mineral fillers in coating system formulations. The examples provide an idea of formulation parameters where fillers exert a direct influence, along with their significance and hints on interpretation. Users will also learn how to select the optimum fillers for imparting desired properties in a given application. Numerous worked examples and practical illustrations round out the discussion on filler applications.

6.1 Importance of fillers in paints and coatings

Fillers were once commonly perceived as nothing more than cheap material for bulking up profits, their sole purpose being to reduce the manufacturing cost of coatings. How times have changed! Users of fillers have since discovered additional, technical functionality in these unprepossessing materials, not least because filler development activity has made progress despite the nay-sayers. There is indeed a great deal of ongoing research and development in the filler field, directed at imparting extra functionality or enhancing established features – see also Chapter 7.

Yet existing, tried and true fillers have a major part to play in paint and coating formulations. Fillers can be pivotal to a variety of key technical properties:

- Increased solids content and filling capacity
- Reduced level of volatile organic compounds
- Enhanced optical properties
- Regulation of coating reflectivity
- Enhanced mechanical properties, e.g. strength
- Reinforcement of coating materials
- Rheology control

Depending on the application and the property requirements profile, selecting the right filler is not always easy. Frequently, that means resorting to tried and true formulas. It is unfortunate that fillers are not always tested and evaluated with the same diligence applied to other classes of raw materials. An oft-heard argument here is that fillers, after water, are the cheapest formula ingredients and therefore do not merit thoroughgoing tests. Accordingly, binders, pigments and additives tend to attract higher priority – this despite the fact that fillers comprise the dominant class of raw material in formulations for interior emulsion paint, façade paints, renders, plasters and other thick film systems, marking paints, primers, and many more besides, see also Chapter 1.2. Such products frequently contain 30 to 40 % filler, rising as high as 80 % in certain applications. It follows that fillers can also play a governing role in the overall property requirements profile of a coating material, and this should not be underestimated.

Meanwhile, natural fillers have attained levels of quality where users may easily forget that these are, in fact, products of nature. It is well known that natural products are subject to variations in their composition, and these subsequently affect the properties of the coating materials into which they are made. Producers of fillers are aware of this, and use selective mining techniques to minimise such variations. But to eliminate them altogether would be unrealistic, so at this juncture it needs to be pointed out once and for all: natural fillers are variable products.

6.2 Important formulation parameters

Special formulation parameters exist, which are important for characterising coating materials that use filler ingredients. Formulation parameters make it possible to classify coating materials, as well as predict the property profile of a given formula.

6.2.1 Non-volatile matter

Solids, solids content (SC) or stoving residue are frequent terms for describing "non-volatile matter" (NVM). ISO 4618 defines what is considered non-volatile matter in coating materials, binders and other raw material ingredients, while ISO 3252 covers the quantitative determination aspects. Non-volatile matter is determined analytically as a means to deducing the composition of an unknown material. Manufactured materials are subjected to similar analysis for quality control purposes. The temperature and drying time used are major factors in the stated result: one popular determination method for non-volatile matter involves exposure to a temperature of 105 °C for 2 hours. The initial weight of the material under test is divided by the output weight after drying and multiplied by 100, see Equation 6.1.

$$NVM = \frac{Weighing\ back}{Initial\ weight} \cdot 100\,\%$$

Equation 6.1: Calculation of non-volatile matter from empirically determined values

For material of known composition like a laboratory test formulation, it is also possible to determine non-volatile matter by arithmetic means alone, without prior measurement.

The arithmetic method requires knowledge of the NVM in each raw material ingredient of the formulation. Figure 6.1 shows a worked example based on a simplified formulation for demonstration purposes. Starting with the NVM in each individual ingredient used and the ingredient's percentage by weight of the overall formulation, it is then possible to calculate the NVM of each formulation ingredient, and from that, the cumulative NVM. Our sample formulation contains 57.6 % NVM of which over two thirds, 41 % to be exact, is attributable to the filler content.

It is fillers that give coating materials their "body", so it is logical that they are major contributors of non-volatile matter. It is possible to reduce the amount of volatile matter present by substituting fillers that have higher oil absorption by others which are less absorbent, with low binder consumption. The result is higher filler and solids content, yet viscosity holds steady. Although very high levels of non-volatile matter are desirable from an environmental viewpoint, there are practical upper limits, due by the requisites of the production process and the way the product is designed to behave when applied to a substrate.

6.2.2 Spreading rate

Spreading rate according to ISO 4618 part 1 is understood to mean the average substrate area that can be covered by a given volume or mass of coating material. Practically speaking, the method of application needs to be capable of producing a single layer in one pass. The resultant spreading rate is stated either in m^2/l, or m^2/kg. Spreading rate is often considered in tandem with coating material opacity. EN 13 300, for example, describes clas-

Function	**Substance**	**NVM in substance [%]**	**Weight [kg]**	**NVM in formula [%]**
Binder	Styrene-acrylic emulsion	50	160.0	8.0
Pigment	Titanium dioxide	100	80.0	8.0
Fillers	Calcium carbonate	100	300.0	30.0
	Talcum	100	80.0	8.0
	Precipitated aluminium silicate	100	30.0	3.0
Additives	Thickener, defoamer, wetting and dispersing agents, preservatives, etc.	33	18.0	0.6
Solvent	Water	0	332.0	0.0
Total			1,000.0	57.6

Figure 6.1: Worked example of calculating non-volatile matter in a simplified emulsion paint formulation

sification of coating materials by their optical and mechanical properties. The practical method for determining opacity measures the contrast ratio between films of differing thickness; they are dried and subsequently weighed to determine their mass per unit area. The result is convertible to spreading rate by factoring-in non-volatile matter and the coating material density. A plot of spreading rate vs. contrast ratio allows interpolation and comparison of contrast ratios for various paints that have a similar nominal spreading rate: 7.5 m²/l in the example shown in Figure 6.2 [1].

Filler content can exert a marked influence on spreading rate in pigmented coating materials. As the filler content increases, so does the solids content and, as a rule, spreading rate as well. These observations only apply where the pigment content is comparable in each case: if the pigment level changes, the spreading rate will change too. Fillers of differing density produce a similar effect, likewise changing the spreading rate expressed in terms of area per unit volume. High-density fillers reduce the volume spreading rate more than lightweight fillers, so coating consumption increases as a result.

6.2.3 Pigment volume concentration

Pigment volume concentration (PVC) is a key parameter of paints and coatings. Along with a few other parameters, PVC makes it possible to predict numerous interrelationships between a coating system's composition, and the resulting properties. PVC makes

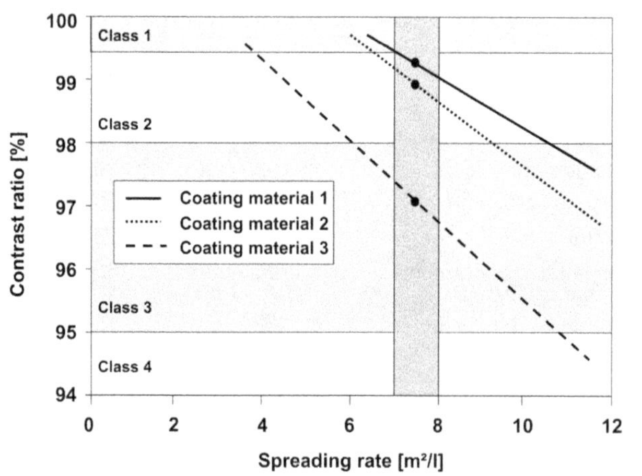

Figure 6.2: Graphic example of correlation between spreading rate and contrast ratio

an equally effective tool for characterising and classifying paints and coatings. ISO 4618 part 1 defines PVC as the ratio of pigment and filler volume in a coating film to the aggregate volume of non-volatile matter. This further implies that the binder is considered in its final, solid form, rather than as supplied. Equation 6.2 shows the mathematical definition of PVC.

$$PVC\,[\%] = \frac{\sum V_{Pigments} + \sum V_{Fillers}}{\sum V_{Pigments} + \sum V_{Fillers} + \sum V_{Binders}} \cdot 100 = \frac{\sum V_{Pigments} + \sum V_{Fillers}}{\sum V_{Total}} \cdot 100$$

Equation 6.2: Definition of pigment volume concentration (PVC)

To illustrate the theory, the following is a PVC calculation example based on the numbers in Figure 6.1. Calculating PVC requires details of the individual raw material ingredients and the masses used, their non-volatile matter and density. Mass is divided by the material's density to convert the mass of raw material to its corresponding volume, see Equation 6.3.

$$\rho \left[\frac{kg}{l}\right] = \frac{m}{v} \Rightarrow V\,[l] = \frac{m}{\rho}$$

Equation 6.3: Conversion of raw material mass to volume

For each of the raw material ingredients involved in the worked example, Figure 6.3 lists NVM, mass as supplied and the NVM component of that mass, plus density. The last column shows the calculated volume of each raw material ingredient.

Once the respective raw material ingredient volumes have been calculated, the PVC can be derived as shown in Equation 6.4. The volumes used in the example result in PVC of 69.6 %.

$$PVC = \frac{(20.0 + 154.5)\ [l]}{(20.0 + 154.5 + 76.2)\ [l]} \cdot 100\,\% = \frac{174.5\ [l]}{250.7\ [l]} \cdot 100\,\% = 69.6\,\%$$

Equation 6.4: PVC calculation for a simplified demonstration emulsion paint formulation based on the data provided in Figure 6.3

As mentioned previously, PVC by itself is already a highly significant parameter. Yet something more is required for a really meaningful interpretation: the critical pigment volume concentration (CPVC)[2]. The relation of PVC to CPVC allows qualitative statements about coating materials and their properties, see also Chapters 6.2.4 through 6.2.6.

Equation 6.2 and the worked example reveal just how strongly PVC is affected by the type and amount of filler used. With coating materials, switching from one mineral filler to another without accounting for their density difference can have drastic results, see Chapter 6.4.1.

Substance	**NVM in substance [%]**	**Weight [kg]**	**NVM in weight [kg]**	**Density [kg/l]**	**Volume [l]**
Styrene-acrylic emulsion	50	160.0	80.0	1.05	76.2
Titanium dioxide	100	80.0	80.0	4.00	20.0
Calcium carbonate	100	300.0	300.0	2.70	111.1
Talcum	100	80.0	80.0	2.75	29.1
Precipitated aluminium silicate	100	30.0	30.0	2.10	14.3
Thickener, defoamer, wetting and dispersing agents, preservative, etc.	33	18.0	6.0	neglected	
Water	0	332.0	0.0	1.00	0.0
Total		1,000.0	57.6		250.7

Figure 6.3: Worked example of calculating the volume of individual raw material ingredients in a simplified demonstration emulsion paint formulation

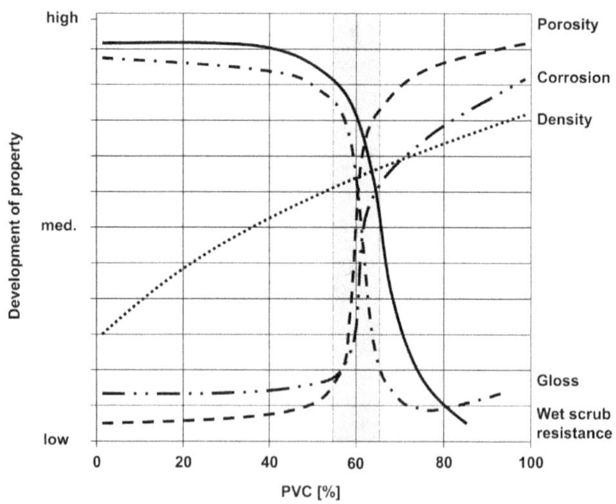

Figure 6.4: Changes in coating properties vs. PVC around the CPVC region

6.2.4 Critical pigment volume concentration

ISO 4618 part 1 covers not just pigment volume concentration, but critical pigment volume concentration (CPVC) as well. CPVC is defined as the PVC at which binder just fills the voids between adjacent solid particles of pigment, filler, etc. in a coating film. Above this point, a number of properties experience significant change, see Figure 6.4. Various properties are predictable based on a coating material's CPVC and PVC, see also Chapter 6.2.5.

To facilitate understanding, Figures 6.5 through 6.7 provide visual representations of three possible PVC states. To simplify the subsequent descriptions, "pigments and fillers" will be collectively referred to as "pigment". Figure 6.5 represents the situation when the PVC lies below the CPVC. The excess binder present (plus wetting agents and dispersants) completely envelopes the pigment surfaces and fills the voids between particles. In this case, the excess binder can produce a glossy film coating. The porosity of such a coating would be zero in theory. In practice, microfoam might form and make it very slightly porous. Evidence of this phenomenon can be obtained by mercury porosimetry [3].

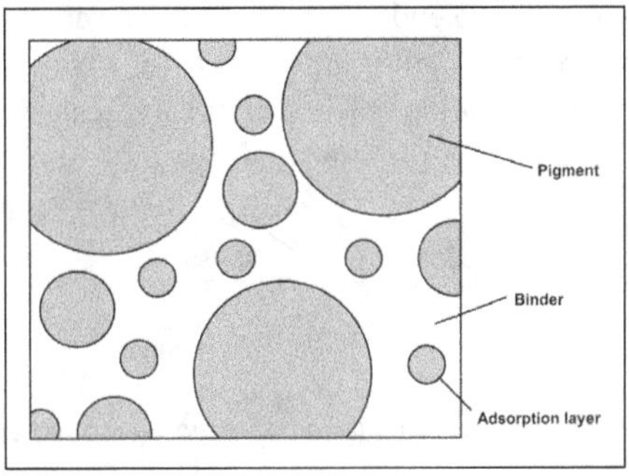

Figure 6.5: Schematic representation of conditions when PVC < CPVC

As the pigment content or PVC rises towards the CPVC, the free binder volume steadily falls and pigment particles approach correspondingly closer. At the CPVC there is no excess binder remaining, so pigment particles begin to contact each other. At this point, the pigment surfaces are still completely wetted with binder, wetting agent and dispersant (adsorbent layer). With the adsorbent layer to keep the pigment particles slightly apart, and binder

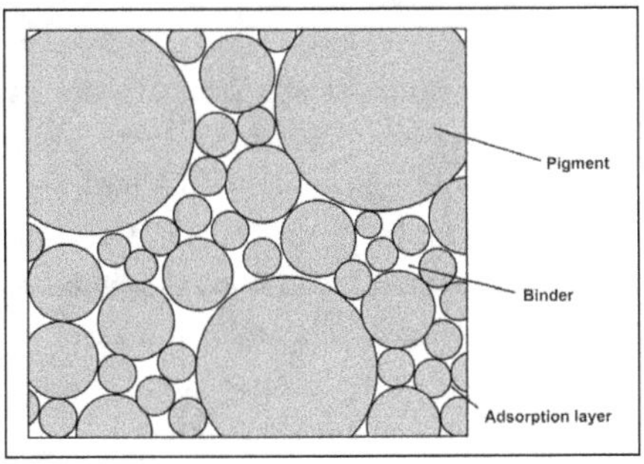

Figure 6.6: Schematic representation of conditions when PVC = CPVC

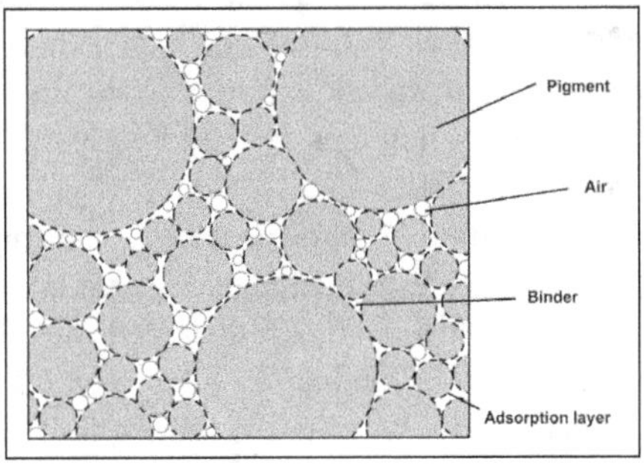

Figure 6.7: Schematic representation of conditions when PVC > CPVC

still completely filling the residual voids between them, porosity is theoretically zero, see Figure 6.6. Coating films in this state exhibit a matt, closed film surface.

When the PVC rises further still, the coating film lacks sufficient binder to fill the voids between pigment particles. A coating film in an over-critical state, i.e. above the CPVC, has porous air inclusions. Rising porosity or PVC causes a steady decline in properties like mechanical strength. In this state, the porous coating film can no longer be weather-resistant because there is insufficient protective binder to enclose the pigment surfaces. The pigment is thus directly exposed to UV radiation, with results similar to those observed in a weathered coating film on a building façade where chalking is already apparent. For this reason, over-critically formulated coatings are recommended for interior applications only. Figure 6.7 shows a schematic representation of a coating film above the CPVC.

The CPVC point can be calculated theoretically, but the task grows progressively difficult as the amount of pigment (P) and fillers (F) in the coating material increases. The difficulty arises because calculation requires the density and oil absorption of the pigment/filler blend, and oil absorption can only be estimated. An overall approximation here is prone to significant errors if the pigment and filler particles are of dissimilar size and shape. Equation 6.5 illustrates CPVC calculation from oil absorption (OA) of the P/F blend, its density (ρ_P) and the density of the binder (ρ_B).

$$CPVC\,[\%] = \frac{100\,\%}{1 + \frac{\rho_P}{\rho_B} \cdot \frac{OA}{100}}$$

Equation 6.5: Calculation of critical pigment volume concentration (CPVC)

CPVC can also be determined by empirical means. The two methods in most common use work by measuring film tension and film porosity respectively over a given PVC range. The film tension method relies on the fact that tension peaks at the CPVC point. A strip of thin material, PVC foil for example, is coated on one side and left to dry. Film tension develops as the coating dries, causing the foil to warp. The amount of warping indicates the degree of film tension, with the CPVC at the point of maximum warping. Test results are reproducible with an accuracy of approx. 1 % [4]. The example in Figure 6.8 indicates the CPVC at 60 %, as evidenced by maximum warping of the single-side coated PVC foil.

The film porosity method for determining CPVC makes use of the physical phenomenon described earlier in this chapter, whereby porosity develops and steadily increases as the PVC rises beyond the CPVC. Porosity is made visually apparent by application of a tar solution (Gilsonite) onto the coating film, which leaves a permanent stain where it is absorbed. The onset of staining marks the CPVC. By measuring whiteness before (R_y white) and after (R_y Gilsonite) absorption via the white substrate of the contrast card, film porosity can be accurately determined and calculated according to Equation 6.6. The more intense the stain, the higher the film porosity. Absence of visible staining means that the film is non-porous, with PVC below CPVC. Although measurement of film porosity with Gilsonite solution is a reproducible method, it can prove less dependable with very densely packed coating films: here, the pores are so minute or inaccessible that they cannot fully absorb the Gilsonite solution. Of the two methods described, film tension is therefore the "safer" way to determine CPVC.

$$Film\ porosity\left[\%\right] = \frac{R_{y_{white}} - R_{y_{Gilsonite}}}{R_{y_{white}}} \cdot 100$$

Equation 6.6: Calculation of film porosity to determine critical pigment volume concentration (CPVC)

Coating material CPVC, and optical properties, all depend very much on the type and quantity of filler used. Pigments and binders are further contributory factors that will not be discussed here [4]. In the Gilsonite method, differences between fillers are expressed as the change in brightness reference value (ΔBRV = R_y white - R_y Gilsonite) vs. PVC. The change becomes more pronounced with increasing porosity and PVC. Below the CPVC, ΔBRV is less than 3. Figure 6.9 shows how various fillers with differing mineralogy, particle size and oil absorption behave over a broad PVC range. Precipitated aluminium silicate A, with very high oil absorption, is the first to show a rising change in brightness reference value. It follows that precipitated aluminium silicate has the lowest CPVC, at around 40 %. It takes quite a while before the coarser carbonates start to show a rise, and this corresponds with their respective oil absorption.

Comparable results are obtainable from empirical determination of CPVC by peak film tension. Figures 6.10 through 6.12 clearly show the varying effects of film tension on

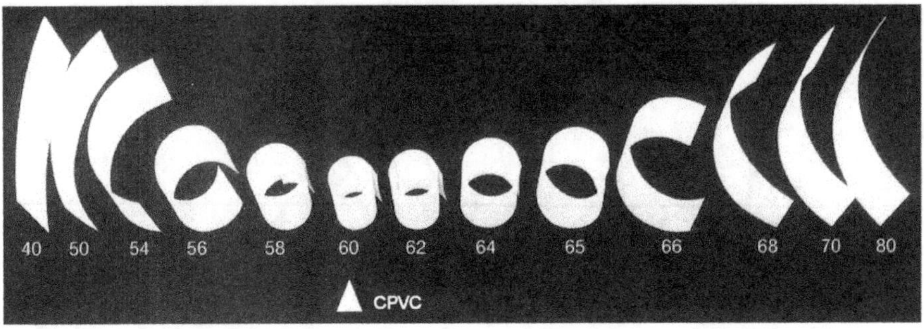

Figure 6.8: CPVC determination from film tension of coating materials with differing PVCs applied to one side of foil strips

single-side coated PVC foil. Maximum warping corresponds with the CPVC of the coating system under test.

To conclude this broad discussion, it needs stating that CPVC can vary greatly as a consequence of the filler and pigment composition. In the example chosen, very fine-particle fillers with high oil absorption (precipitated aluminium silicate A) result in a CPVC of approximately 40 %. On the other hand, CPVC lies above 65 % in the case of very coarse products with extremely low oil absorption (dolomite A). It is important for developers to appreciate that their coating materials should never be formulated with PVC in the region of the CPVC. Properties of coating materials and films can undergo radical and steep changes around this point, such that a minor upward or downward deviation in PVC can have dramatic consequences. To cite just one final example in this regard: at the CPVC, film tension may be so high that a critically-formulated emulsion paint applied to the front, visible surface of ingrain wallpaper will overcome the strength of the wallpaper paste on the back, causing the paper to detach from the wall! To avoid this and other undesirable effects, the PVC of the coating material should be chosen so as to lie further than ±5 % away from the CPVC.

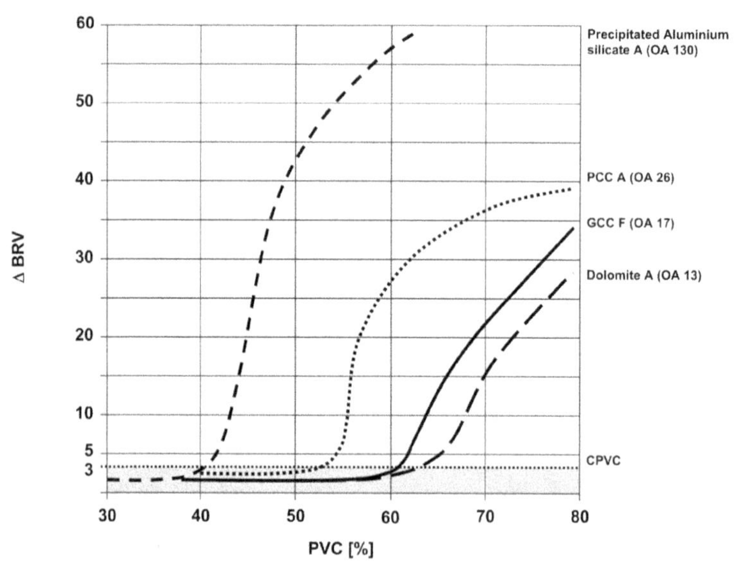

Figure 6.9: Effect of fillers on CPVC as determined by the Gilsonite film porosity method

6.2.5 Pigment/filler loading

The PVC and CPVC parameters described previously are each predictors of coating material behaviour in their own right. They can also be expressed in a combined form as their quotient, or Q-value, see Equation 6.7.

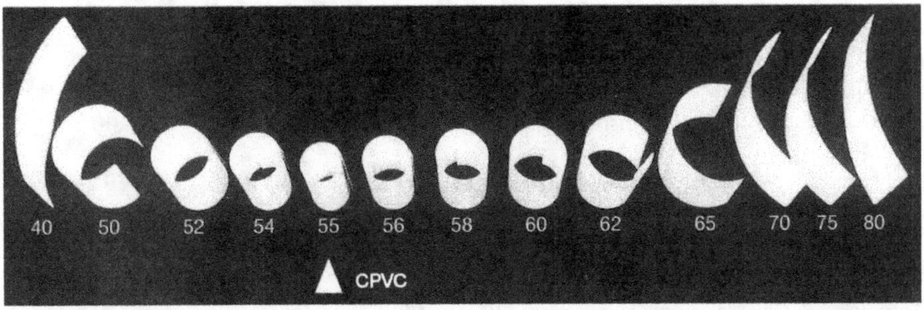

Figure 6.10: Effect of precipitated calcium carbonate A (PCC A) with OA of 26 on CPVC/film tension source: Kronos International, Inc.

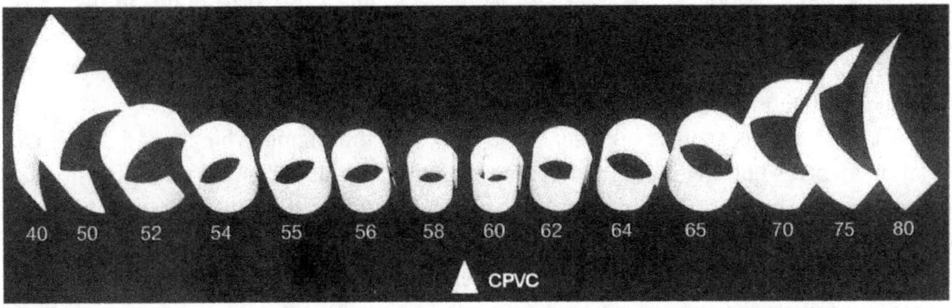

Figure 6.11: Effect of natural calcium carbonate F (GCC F) with OA 17 on CPVC/film tension source: Kronos International, Inc.

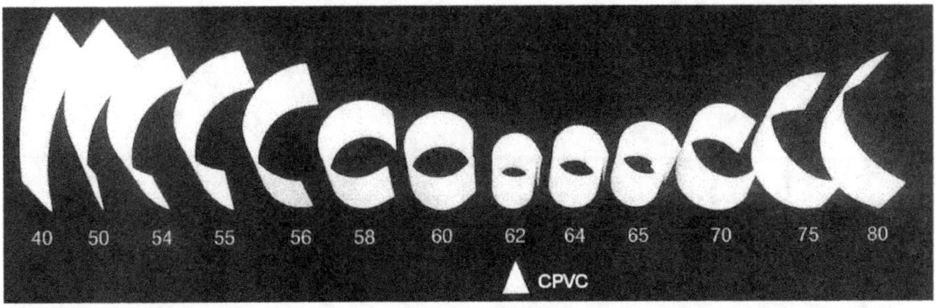

Figure 6.12: Effect of dolomite A with OA 13 on CPVC/film tension source: Kronos International, Inc.

Table 6.1: Coating materials and their Q-values

Q-value	Coating material	PVC range [%]
< 0.5	High-gloss topcoats Emulsion varnishes	< 25
0.4 to 0.8	Intermediate coats Primers	20 to 50
0.5 to 0.8	Satin matt and matt topcoats	30 to 50
0.7 to 0.95	Anti-corrosion primers	40 to 60
0.8 to 1.2	Knifing fillers and putties	50 to 80
1.0 to 1.2	Powdered zinc primers	70 to 80
1.05 to 1.3	Interior emulsion paints	70 to 90

$$Q\ value = \frac{PVC}{CPVC}$$

Equation 6.7: Calculation of Q-value from PVC and CPVC

The benefit of this approach is to classify coating materials by a single number. Table 6.1 shows a variety of applications classified by their Q-value. Q=1 corresponds to the CPVC state. Coating materials with a Q-value less than 1 are under-critically formulated, and thus essentially non-porous. Formulations with a Q-value above 1 are over-critically formulated, and hence porous.

Another way to express pigment/filler loading is the pigment/binder ratio. This calculation involves nothing more than the mass of pigment plus filler (specified for simplicity as m_P), and the mass of binder (m_B). It completely ignores density differences between raw materials, and their respective volumes. Coating material specifications frequently state pigment/filler loading in terms of pigment/binder ratio because it is so easy to calculate, see Equation 6.8.

$$Pigment - binder\ ratio = \frac{m_P}{m_B}$$

Equation 6.8: Calculation of pigment/binder ratio

6.2.6 Packing density

Certain packing density states have already received extensive coverage in Chapter 6.2.4, which focused on the case of critical pigment volume concentration. To complete the picture, further details will be discussed here. Interestingly, packing density is an indicator of various properties of coating films. This discussion uses representative schemes for simplified explanatory purposes.

Fillers exhibit differing widths of particle size distribution. This width, along with the shape of the particles themselves, extensively influences packing density in coating films. Fillers on the whole match one of three possible schemes:

- Uniform particle size
- Narrow particle size distribution
- Broad particle size distribution

A broad particle size distribution may also be the result of using a filler blend. Figures 6.13 through 6.15 schematically represent the situation for these three particle size distributions in the case of nodular (spheroid) fillers. In the representative broad particle size distribution, the smallest particle measures 0.5 µm and the largest 5 µm. In practice, real fillers are quite likely to contain particles ranging in size from 0.1 to 45 µm. The figures show wide gaps between particles, which in a coating film would be filled by excess binder.

Reducing the amount of binder decreases the volume of the coating film, so filler particles come into contact. The maximum-density packing for fillers with uniformly sized particles lies at above two-thirds, 74 % to be exact. Air, or binder, fills the residual volume of the coating film. Figure 6.16 represents the situation in two dimensions.

A filler with a narrow to broad particle size distribution that is packed at the highest achievable density will exhibit even less residual space between particles. As the two-dimensional representations in

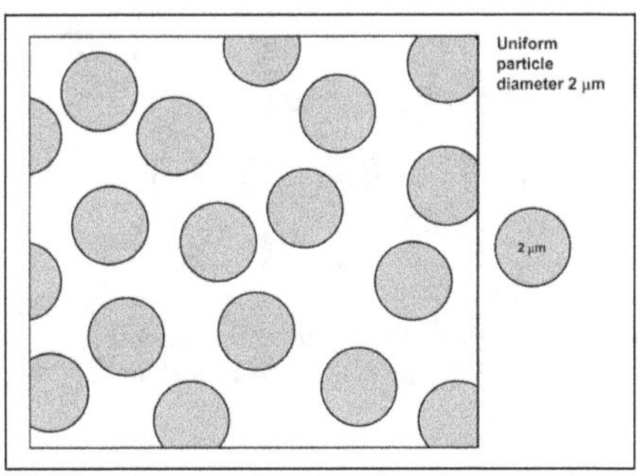

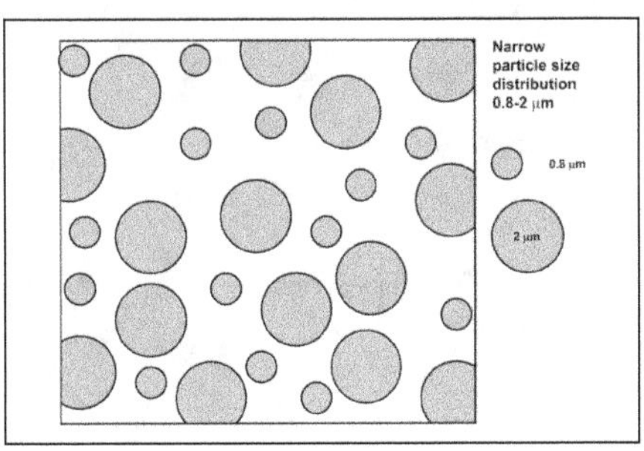

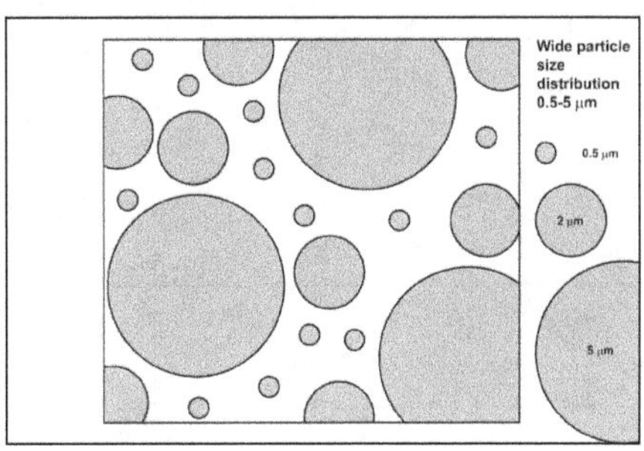

Figures 6.13 through 6.15: Coating films with fillers having uniform particle size (6.13), narrow (6.14) and broad (6.15) particle distribution

Figures 6.17 and 6.18 show, the packing density steadily increases as the particle size distribution grows wider.

Disregarding exceptions like polymer hollow spheres, it is rare indeed for practical fillers to exhibit even remotely uniform particle size distributions. Instead, the vast majority more closely resemble the broad particle size distribution model. Figure 6.18 indicates less residual space than the scheme of Figure 6.17, which is in turn considerably less than the residual space indicated by Figure 6.16. Consequently, CPVC is higher in very densely packed coating systems than in less densely packed ones. Practical formulations for anticorrosion primers [3], smoothing fillers, putties and similar applications use exactly this principle to reduce theoretical porosity and hence capillaries that might produce diffusion. A blend of differently shaped particles (lamellar mixed with nodular) results in greater porosity than if the particles were all the same shape. Maximum-density packing is achieved by using nodular fillers with an extremely broad particle size distribution.

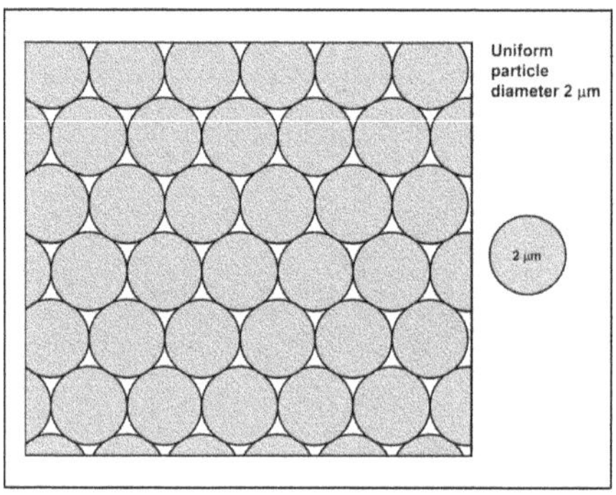

Figure 6.16: Coating film with maximum-density packing of filler containing uniformly sized particles

Although maximum-density packing can bring a variety of beneficial effects in a coating film, it is not always optimal from the application viewpoint. In very high solids systems especially, dense packing can be a cause of processing problems: the high packing density restricts particle mobility and thus increases viscosity. Extreme cases can even result in shear thickening, or dilatancy, which may also prove a hazard in the production of coating materials.

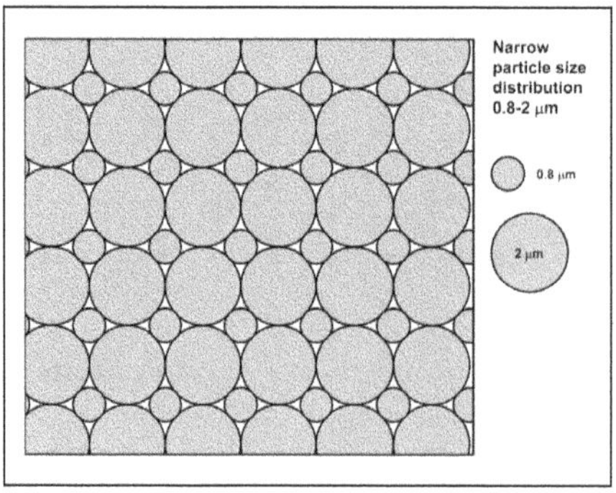

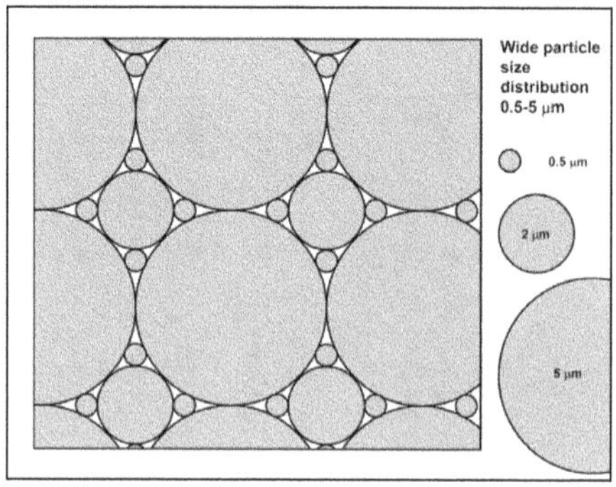

Figures 6.17 through 6.18: Coating films showing maximum-density packing of fillers with a narrow (6.17) and broad (6.18) particle size distribution

6.3 Filler influences on coating materials

Fillers can have a great many influences on wet coatings. Users of fillers will frequently find that the resulting properties are not precisely predictable; surprises always occur. The following chapters examine a variety of typical properties and observations related to fillers in wet coatings.

6.3.1 Dispersibility

There are simple and obvious reasons why easily dispersible fillers are such an important production criterion for manufacturers of paints and coatings. Not only does poor dispersibility make dispersion a longer process, but more complex dispersing devices may be necessary as well. A time and capital-intensive dispersion process implies higher production costs. Yet apart from these commercial aspects, poor or unsatisfactory dispersibility can make it impossible to achieve a finished product with the required properties. Visible symptoms on the coating film can include impaired and less durable gloss, opacity and colour strength, or in extreme cases a peppery surface.

So just what happens during filler dispersion? It initially helps to consider the filler in isolation. Whatever the filler production process – grinding or chemical precipitation – primary particles are the general result. But as these make their way through the produc-

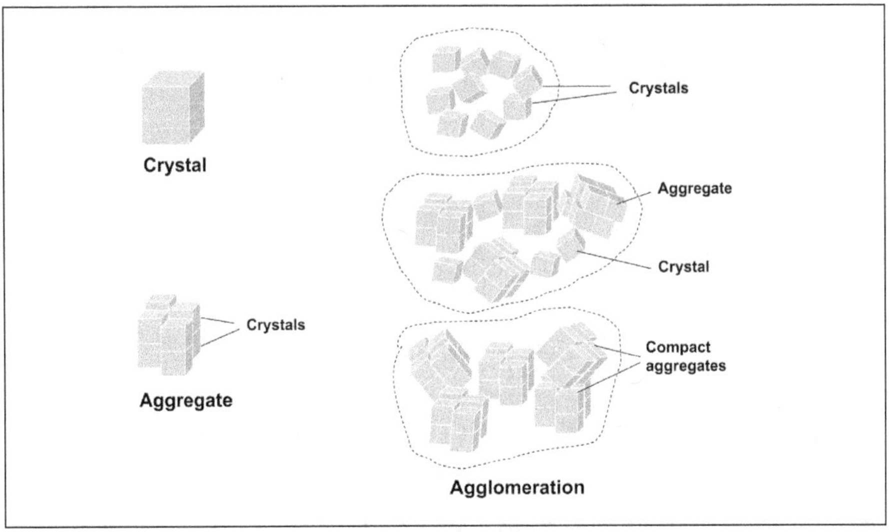

Figure 6.19: Filler morphology, showing crystals, aggregates and agglomerations

tion chain, they tend to form aggregates or agglomerations. Aggregates consist of primary particles with abutting surfaces, and exhibit less overall surface area than their constituent primary particles. An agglomeration is defined as a collection of primary particles and/or aggregates joined by their corners and edges. Its overall surface area does not differ significantly from the cumulative individual surface areas of the constituent primary particles or aggregates, see Figure 6.19. Particle morphology is defined in DIN 53206.

There are various ways of testing filler (and pigment) dispersibility, and these are described in ISO 8780 parts 1 through 6. The standard includes methods for dispersion machinery like bead mills, oscillating shakers and triple roll mills. These break up agglomerations and to some extent aggregates as well by applying heavy shear forces. Using "only" a dissolver puts greater demands on the filler's dispersibility, given the more modest shear forces that dissolvers are capable of producing. It follows that dispersion tests are most critical in association with dissolvers. Dissolver tests involve dispersion under conditions defined in ISO 8780 part 3, followed by determination of grinding fineness using a grindometer wedge. ISO 1524 describes the method for determining grinding fineness, which is applied after a specific dispersion time. Further methods to track the state of dispersion include recording gloss or colour strength development over time, see ISO 8781. The dispersion results for the solvent-containing alkyd enamel shown in Figure 6.20 were obtained after a dispersion time of 8 minutes at a dissolver disc peripheral speed of 9.5 m/s.

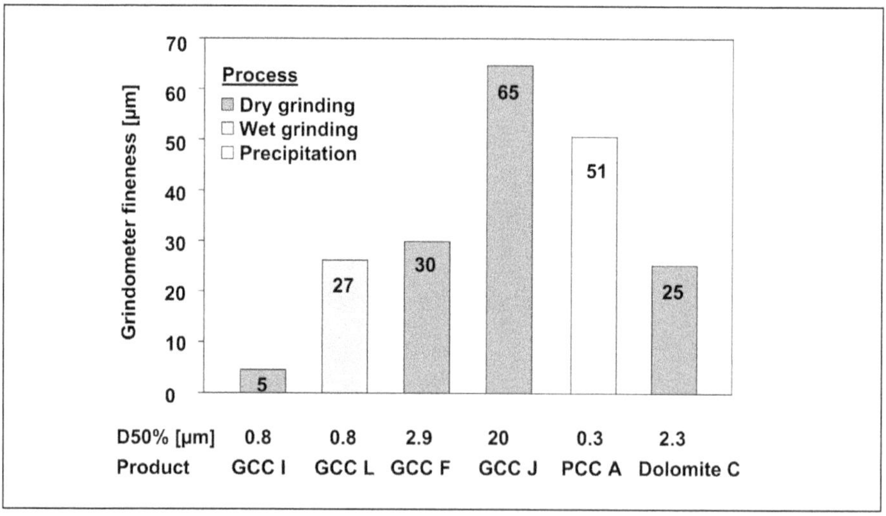

Figure 6.20: Grindometer determination of dispersion fineness of fillers produced by various means, with differing median particle diameter D50%; 8 minutes' dispersion time in solvent-containing alkyd enamel at 9.5 m/s peripheral speed

The results for dispersion fineness show some major differences. It is clearly apparent how median particle diameter D50% does not always permit conclusions about dispersibility. The dispersion results correspond more closely with the largest primary particle size (top cut), or stable agglomerations [5]. Figure 6.21 illustrates dispersibility over time for a variety of fillers as well as titanium dioxide.

Dispersion was performed according to ISO 8780 part 2. Each of the selected fillers has the same median particle diameter, yet there are clear differences of behaviour in the development of grindometer fineness and the end result. Filler dispersibility is considerably influenced not only by fineness per se, but also by the production process: dry or wet grinding, or precipitation.

6.3.2 Rheology

Mineral fillers exert considerable rheological influence on coating materials, via a number of mechanisms. Of rheological significance are:

- Particle size
- Particle structure
- Filler content

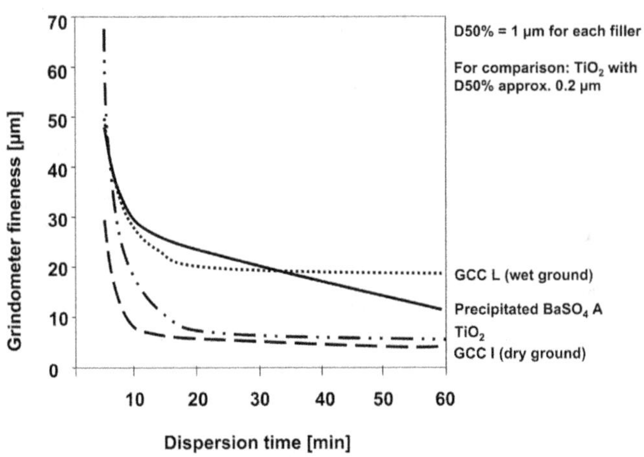

Figure 6.21: Grindometer measurement of developing dispersion fineness in solvent-containing alkyd enamel, for a variety of fillers with the same median particle diameter D50% of 1 µm and produced by different means

Specific surface area and oil absorption both increase as the size of the filler particles decreases. Binder consumption and hence viscosity likewise increase steadily. Particle size distribution is also a factor in filler fineness, and the way that fineness affects viscosity. At a given median particle size, a filler with a wide PSD will generally contain greater numbers of fine particles than one with a narrow PSD. This was the reason for developing fillers that have the fine particles removed by an extra screening process, specially for use in low VOC, high-solids coating materials. Such fillers facilitate increased solids content without the increased viscosity that would result from using filler with a wide PSD. Figure 6.22 illustrates the effect of varying fineness and narrow/wide PSDs on viscosity and filler content, using talcum fillers as an example.

Particle shape affects viscosity for a variety of reasons. One is the relation of particle shape to specific surface area/oil absorption. Nodular fillers exhibit lower oil absorption, and hence viscosity, than lamellar fillers of similar fineness. Particles at rest may interact to form networks and frameworks, both of which tend to increase viscosity. Breaking up such networks and their intrinsic viscosity calls for a certain amount of shear force. The filler particles align themselves with the shear direction, so viscosity decreases, and vice versa.

The effect of filler content on viscosity is already apparent from Figures 6.22 and 6.23, which show viscosity increasing as the filler content goes up. This occurs considerably sooner in fillers with high oil absorption (e.g. kaolin, talcum and mica) than in fillers

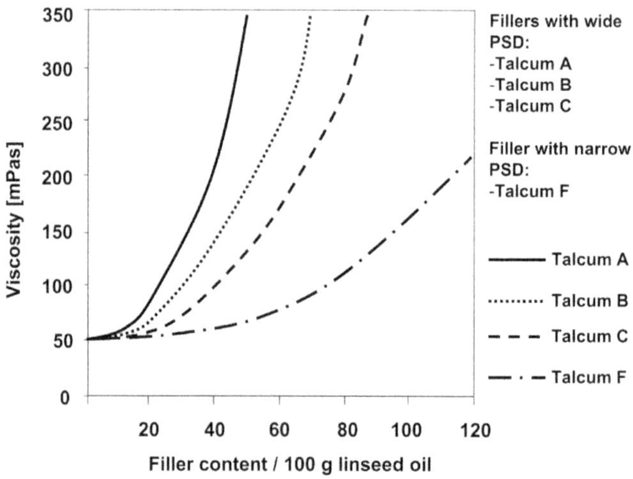

Figure 6.22: Viscosity of various grades of talcum fillers with narrow and broad PSDs versus filler content

with lower binder consumption[6]. Cristobalite, GCC and natural barium sulphate are all examples of fillers with low oil absorption, see Figure 6.23.

In conclusion, fillers act to increase viscosity on account of their particle size and shape, and the filler content. However, that does not mean filler alone is capable of regulating viscosity. Next to the binder, it is mostly organic thickeners, by themselves or in combination, that are responsible for a well-developed rheological profile. Organic thickeners are considerably more effective than regular fillers, the exceptions being very fine pyrogenous silicas and precipitated silicic acids. The latter can indeed make a large rheological contribution. It is not possible for Figures 6.22 and 6.23 to depict the powerful thickening effect caused by their interaction. Using such fillers produces a three-dimensional framework held together by hydrogen bonding. As well as the actual thickening effect caused by pyrogenous silicas and precipitated silicic acids, there exists a strongly time-dependent viscosity function, called thixotropy[7]. The dosing levels required are usually on the order of a few percent.

6.3.3 Wet hiding power

It is well-known that fillers in coating materials contribute little to wet hiding power: their refractive indices are too low, or too similar to that of the binder. This is why large quanti-

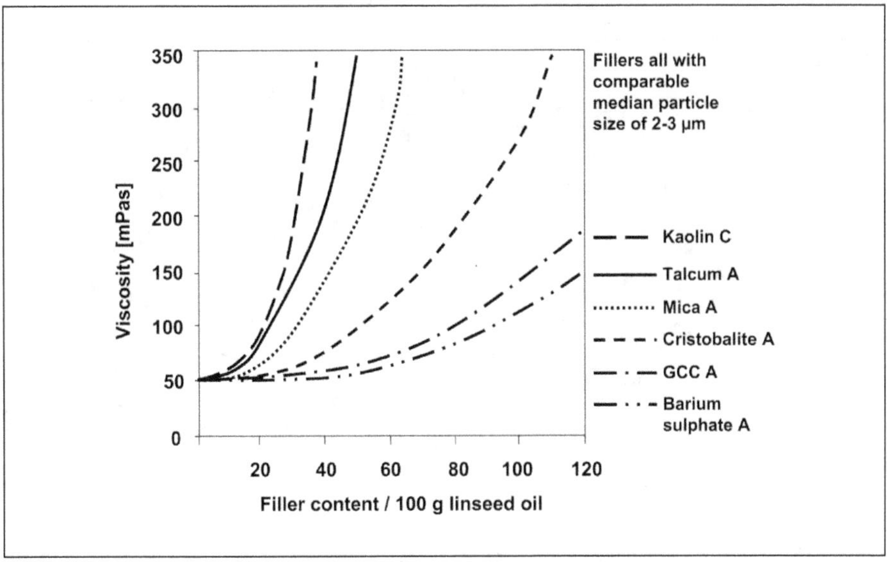

Figure 6.23: Viscosity of various fillers having comparable particle size

ties of white pigment, especially titanium dioxide in its rutile modification, go into under-critically formulated (PVC < CPVC) and binder-rich paints and coatings. Adequate wet hiding power is vital in products intended for manual application, so the user can see which areas have already been painted. Wet hiding power is not an issue in automated application processes, where the machinery is computer-controlled and a visible coating is not essential. But coating materials intended for manual application must all contain sufficient pigment to produce the necessary wet hiding power.

Despite their low refractive indices, fillers do possess wet hiding power to some degree, albeit much less so than pigments. Not all fillers are the same in this regard, and their wet hiding power is tested in unpigmented coating materials with an average PVC. In comparison tests of mineral fillers with differing density, it is important to observe a constant PVC in each case. After the test formulations are made, they are applied to contrast cards (opacity charts) and left to dry. After a suitable interval of 7 days, the cards are irrigated for 30 minutes at 23 °C. The surface is dried off, and R_y whiteness over black and white substrate is measured by a spectrophotometer. Figure 6.24 shows results obtained by this method for a variety of fillers. The formula for testing the fillers is based on a 50 % styrol acrylate emulsion and contains the usual wetting and dispersal, defoaming, thickening, pH regulation and coalescing additives. The PVC of the formulation is 50 %.

With constant formulation parameters, observable differences in wet hiding power range from 20 to 50 %. Differences of this magnitude are astonishing when one consid-

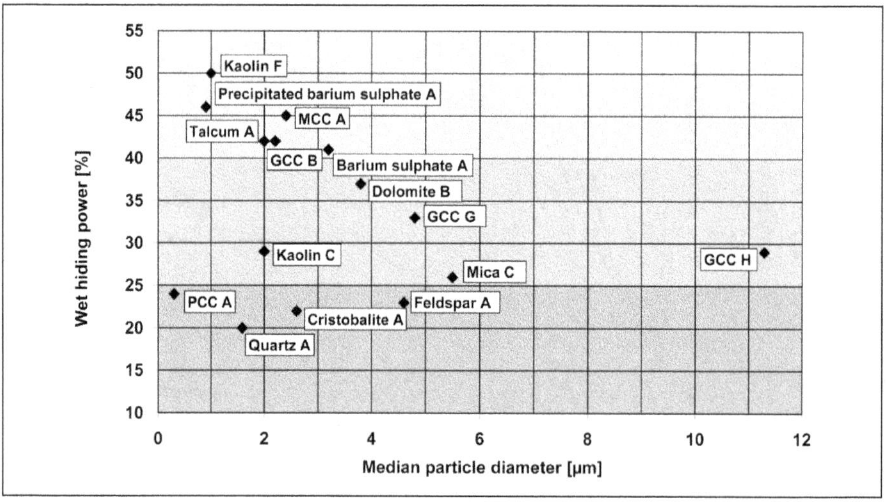

Figure 6.24: Wet hiding power of various mineral fillers, application by draw down bar with an open gap of 150 µm

ers that the refractive indices of kaolin, talcum, barium sulphate and GCC (natural calcium carbonate) are not so different from those of quartz, cristobalite and PCC (precipitated calcium carbonate). It can be seen as a practical remarkable confirmation of theory that fillers with internal air voids (flash-calcined kaolin = kaolin F) achieves the greatest wet hiding power. Considering fillers with the same mineralogy, it is equally apparent that wet hiding power increases with particle fineness (GCC series). Synthetic fillers are not necessarily capable of greater wet hiding power (see calcium carbonate), although this is still quite possible (see barium sulphate). It is more or less impossible to point out one mineralogy that is especially effective in terms of wet hiding power. In our example, barium sulphate fillers are the only ones to maintain a high level, whereas silica fillers achieve only modest wet hiding power.

6.3.4 Storage stability

Coating materials must maintain a satisfactory period of storage stability. This means a minimum of six months to two years, depending on the coating system. A producer of coating systems is expected to deliver the stated shelf life, and this means developing formulations from mutually compatible raw material ingredients. In coating materials, the filler aspects of storage stability largely come down to sedimentation and/or reaction with the binder.
The main factors influencing sedimentation of fillers are:
- Density
- Particle shape and fineness
- Dispersibility
- Stabilisation with the aid of wetting agents and dispersants.

Developers of coating systems can choose among fillers of widely differing density. Low-density mineral fillers are the first choice for reducing the risk of sedimentation. That having been said, barium sulphate fillers are well known as being the densest by far, yet they remain an established feature of paints and coatings, especially for automotive and general industry. Thoroughly stabilising the filler particles with wetting agents and dispersants is essential. Inadequate affinity between the mineral filler particles and organic wetting agents and dispersants jeopardises stabilisation from the outset. As well as affinity (hydrogen bond interactions, hydrophobic/hydrophilic behaviour), molecular weight, the length of the polymer chains, and possible modifications to wetting agents and dispersants play major roles as well. If the polymer chains are too short in relation to the filler surface, there is a risk of fillers forming agglomerations, see Figure 6.25. The polymer chains then cling to the agglomerations. Figure 6.26 shows the opposite situation, where the wetting agent and dispersant polymer chains are over-long and thus create a risk of flocculation. This is why matching the wetting agent and dispersant to the filler (Figure 6.27) is frequently more crucial than the filler's density [8, 9].

Fillers stabilise by two mechanisms: electrostatic stabilisation predominates in water-based coatings, and stearic stabilisation in solvent-containing ones. Should sedimentation occur anyway, a soft, easily stirrable sediment is definitely an advantage. Developers can choose between many wetting agent and dispersant products, which are additives typically based on polyacrylate, polyurethane, polyurea, or a mix of the three. Polyacrylate wetting and dispersing agents are definitely the most popular. Properly chosen wetting agents and dispersants make fillers more easily dispersible and tend to break up aggregates and agglomerations. The existence of these is comparable to the presence of very coarse filler particles, both of which precipitate to the bottom very fast.

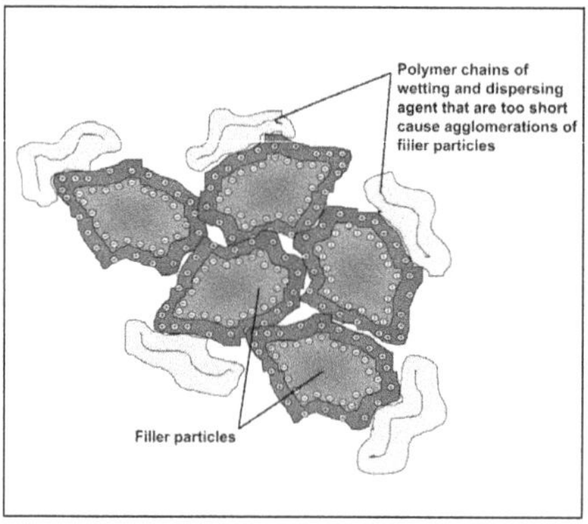

A further effect of sedimentation can be serum formation (clear liquid) on the top surface of the coating material. Specialists call this phenomenon syneresis. It too produces a concentration or phase shift in the coating material composition, and can have a marked effect on storage stability.

Very fine fillers can be most beneficial to stability, given that sedimentation rate is proportional to the square of the particle radius according to Stokes' law. From a physical viewpoint, it follows that simply

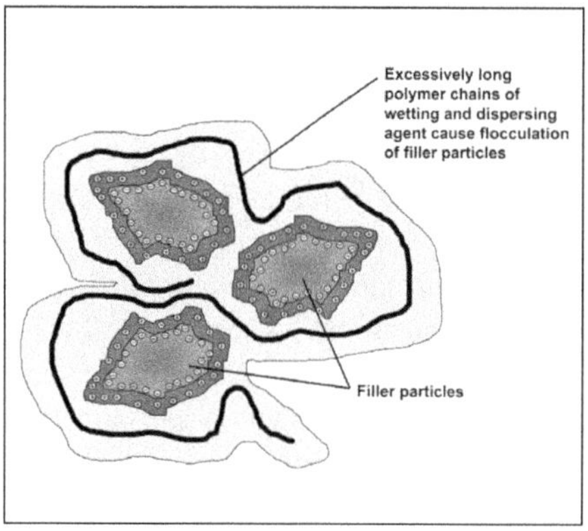

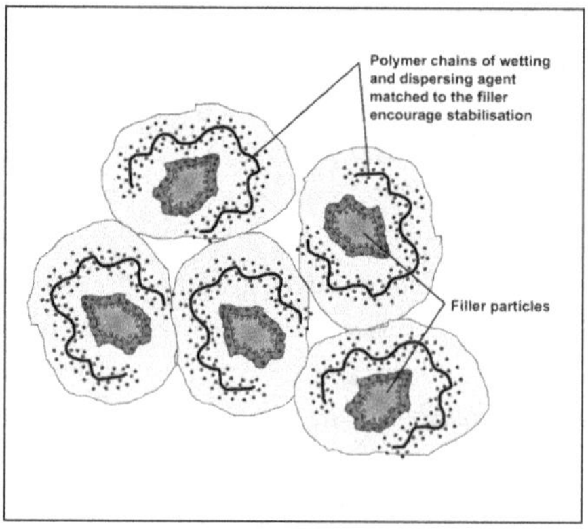

Figures 6.25 through 6.27: Mineral fillers in unstable and stable wetting states – agglomeration (6.25), flocculation (6.26) and stabilisation (6.27) *source: Coatex SA*

using very fine fillers will decrease the sedimentation rate. Moreover, coarse filler particles are very heavy as well, which reinforces the tendency to sedimentation.

Lamellar fillers like talcum, kaolin and mica also tend to discourage sedimentation. At rest, they increase viscosity by forming frameworks, and hence reduce the tendency to sedimentation. Known to have a considerably stronger effect still are pyrogenous silica and other fine-particled fillers with primary particles on the nanometre scale. Their capacity for interaction means they can form three-dimensional frameworks, which produce pronounced thixotropy and thus increase storage stability.

Reactions between fillers and binders or the coating material itself can markedly influence storage stability. Fillers may react when the coating material medium causes them to stop behaving in an inert manner. It is well known, for example, that carbonate fillers have no acid resistance. To use these in cationic systems would make the carbonate turn into its corresponding hydroxide plus carbon dioxide. Further reactions, for example with carboxyl groups, are also possible and may destabilise the entire system by causing a sharp viscosity increase. This is why it is recommended to use only chemically inert fillers in cationic systems. Yet here too, low concentrations of soluble ions from mineral impurities can be problematic in water-based systems, whether the mineral filler is chemically inert or not.

6.4 Filler influences on coatings

Fillers influence the properties of dried coating films just as much as they do the wet coating material before it is applied to a substrate. The relative significance of specific properties depends on the application area. Optical properties like opacity, brightness and reflection are certainly high on the list of priorities, as are mechanical properties like wet scrub resistance, film hardness, adhesion, and resistance to chemicals.

6.4.1 Hiding power

Although mineral fillers have only modest refractive indices, they play a far from insignificant role in imparting hiding power to coating materials. While they contribute somewhat less to wet hiding power than pigments, see also Chapter 6.3.3, they still have a subsidiary function. Fillers' influence on hiding power is largely due to two mechanisms:

- Hiding power caused by the dry hiding effect in over-critically formulated coating materials
- Pigment particle spacing, which gives a subsidiary boost to hiding power.

A coating material's hiding power is defined in ISO 2814 as the percentage contrast ratio of R_y whiteness as measured against a black vs. a white substrate, see Equation 6.9.

$$Contrast\ ratio\ \left[\%\right] = \frac{R_{y_{black}}}{R_{y_{white}}} \cdot 100\ \%$$

Equation 6.9: Formula for ISO 2814 hiding power, expressed as contrast ratio

In over-critically formulated coating materials like emulsion paints, hiding power is largely a function of the following filler properties:
- Particle fineness
- Oil absorption
- Brightness

Particle fineness is predefined by the filler producer; it cannot be altered in the coating material by dispersion and grinding processes. Accordingly, developers must select mineral fillers with fineness appropriate to a given application. Particle fineness correlates more or less with the filler's oil absorption, see Chapter 4.1.2.3. The third property, brightness, is intrinsic to the filler material and is only slightly improvable by selection and washing during the production process. Fillers lacking in brightness always create a subjective impression of greater hiding power, because their limited whiteness prevails even over a white substrate. It follows that the difference in whiteness over black and white substrates is less than with very bright fillers. The result is a higher contrast ratio, albeit with less whiteness.

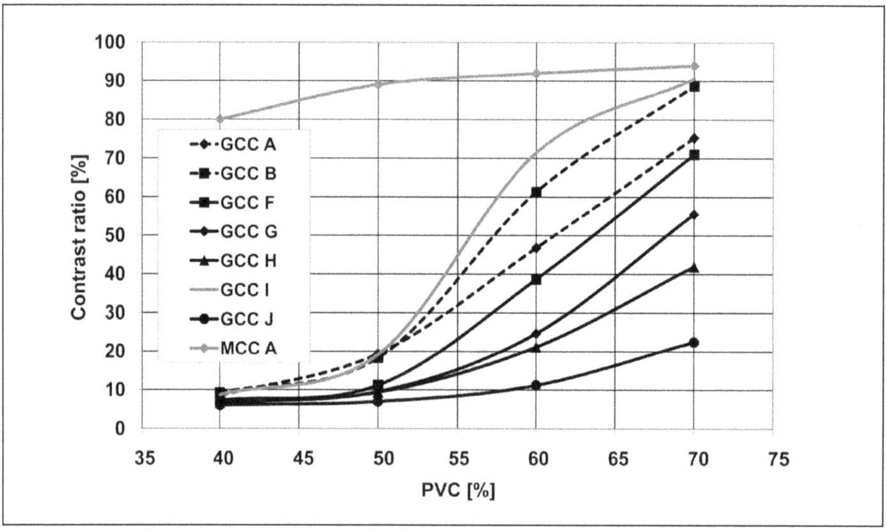

Figure 6.28: Contrast ratio vs. PVC for calcium carbonate fillers of differing fineness

The hiding power examples below repeatedly concentrate on the same criteria: particle fineness and PSD, plus oil absorption/ BET specific surface area.

The examples here are all based on emulsion paint with styrol-acrylate binder, along with the usual thickening, defoaming, dispersant, preservative and coalescing additives. The formulations contain no white pigment, so as to better reveal the filler's own characteristic behaviour. The differences shown will of course be less apparent in practical coating materials that contain white pigment. The paints were all applied to contrast cards using a draw down bar with 150 µm gap, then left to dry for 3 days at 23 °C and 50 % relative humidity before testing their application-relevant properties.

The example shown in Figure 6.28 is typical of emulsion paint fillers with a common mineralogy and production process. Hiding power, expressed in terms of

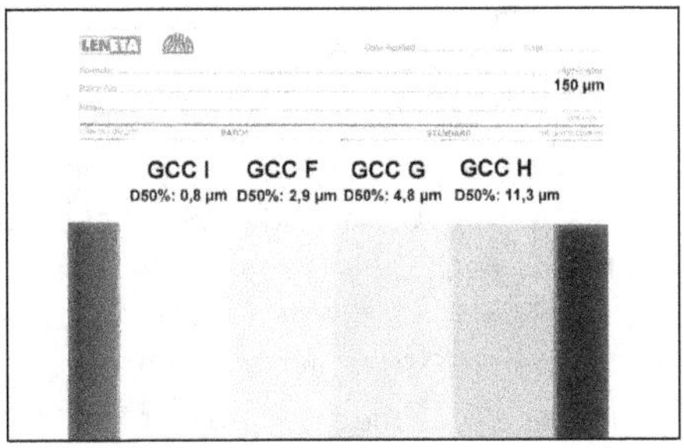

Figure 6.29: Parallel draw downs of calcium carbonate fillers at constant 70 % PVC and varying fineness

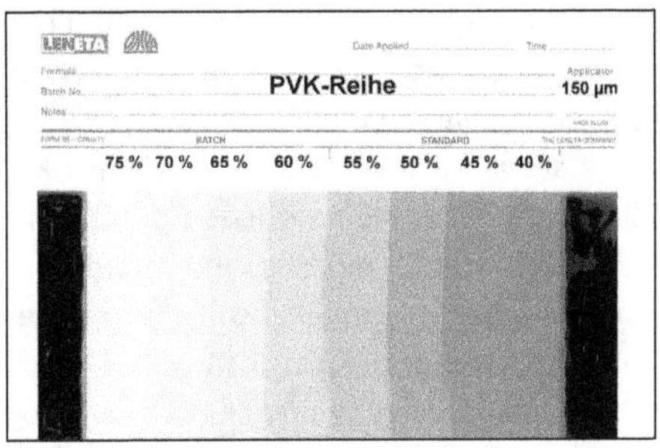

Figure 6.30: Parallel draw down of a 2 µm calcium carbonate filler at various PVCs

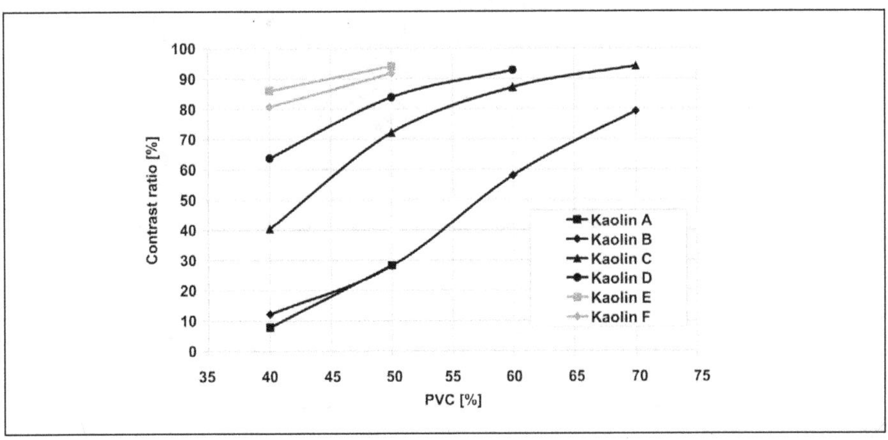

Figure 6.31: Contrast ratio vs. PVC for kaolin fillers produced by different methods

contrast ratio, increases with filler fineness and hence rising oil absorption. The finest fillers shown in Figure 6.28 are GCC B and GCC I; the coarsest is GCC J. At comparable PVC in each case, the hiding power of fillers with a small median particle size starts to rise earlier than is the case with coarser products. In our example, the hiding power of GCC A, B and I at 50 % PVC is already twice that of coarser GCC fillers. The optical performance of fine fillers like GCC I is considerably greater than that of GCC G: GCC I provides the same hiding power at 57 % PVC as GCC G achieves at a PVC of 70 %. Alternatively speaking, GCC I at PVC of 70 % achieves a 90 % contrast ratio, compared to just 55 % with GCC G: in other words the finer product has 35 % more hiding power. To illustrate the origin of these graphically plotted values, Figures 6.29 and 6.30 show contrast cards with paint draw downs. In contrary to GCC, a modified calcium carbonate shows completely different performance. The much higher specific surface area and oil absorption leads at PVC levels of 40 and 50 % already to high film porosity. This results for MCC A in high dry hiding power of approximately 90 % at PVC 50 %, although the mean particle size is clearly coarser compared with GCC A and I.

Natural, calcined and flash-calcined kaolin are examples of mineralogically identical fillers with disparate production processes. Each distinct process, see Chapters 3.1.3 and 5.2.2, significantly modifies the kaolins' properties in terms of particle morphology, specific surface area, and oil absorption. Figure 6.31 shows just how extensive such modifications can be, especially at the low PVC end of the scale. At 40 % PVC, kaolin A and B exhibit a very low hiding power of around 10 %. By comparison, the calcined kaolins C and

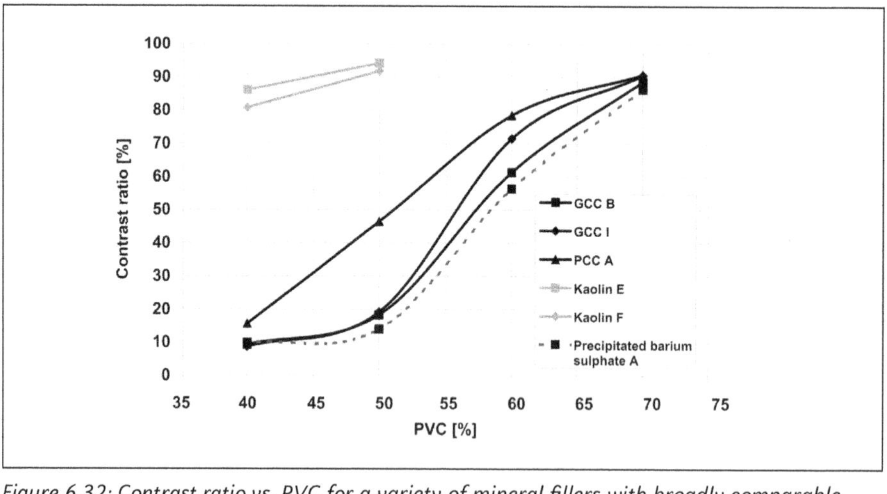

Figure 6.32: Contrast ratio vs. PVC for a variety of mineral fillers with broadly comparable median particle diameters of 1 µm or less

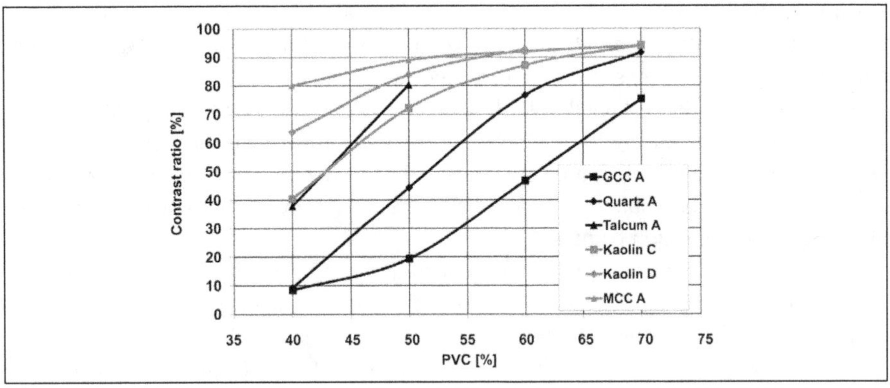

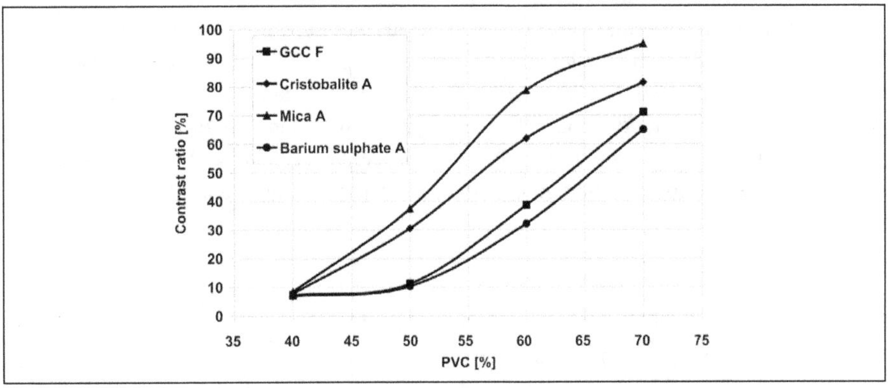

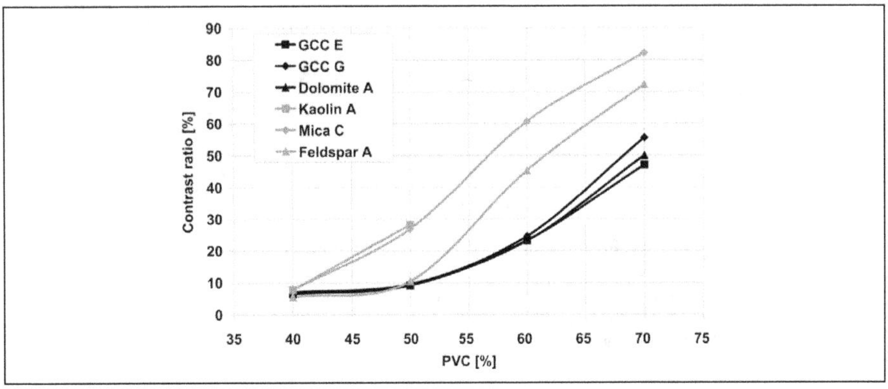

Figures 6.33 through 6.35: Contrast ratio vs. PVC for a variety of mineral fillers with broadly comparable median particle diameters of 2 µm (6.33), 3 µm (6.34) and 5 µm (6.35)

D achieve contrast ratios of 40 and 64 % respectively at the same PVC. The flash-calcined kaolins E and F exhibit yet another marked rise in contrast ratio, to over 80 %. This difference in hiding power comes as the result of calcination and increased binder consumption. The internal air voids contained in flash-calcined kaolins likewise make a positive contribution. Here too, increasing the PVC naturally causes a further increase in hiding power. Flash-calcined kaolins E and F are very strong influencers of viscosity, as is natural kaolin A, such that it is impossible to increase the filler content any further if kaolin is the sole filler used in the emulsion paint.

Figure 6.32 considers a variety of fillers with comparable median particle diameters of around 1 µm or less. The comparative ranking of contrast ratios at given PVC reflects the oil absorption of the respective fillers: kaolin F has the greatest hiding power and oil absorption of 67 g/100 g; the oil absorption of mid-ranked PCC A is 26 g/100 g, while barium sulphate is lowest of all at 13 g/100 g.

As Figure 6.32 shows, the hiding power achievable by fillers of comparable fineness broadly corresponds to their binder consumption, expressed in terms of oil absorption. However, not all fillers are available with a median particle diameter of 1 µm; many products are available with median particle diameters no finer than 2, 3, or even 5 µm. Taking account of this fact, Figures 6.33 through 6.35 provide an overview of the optical performance of coarser products, analogous to the fine example in Figure 6.32.

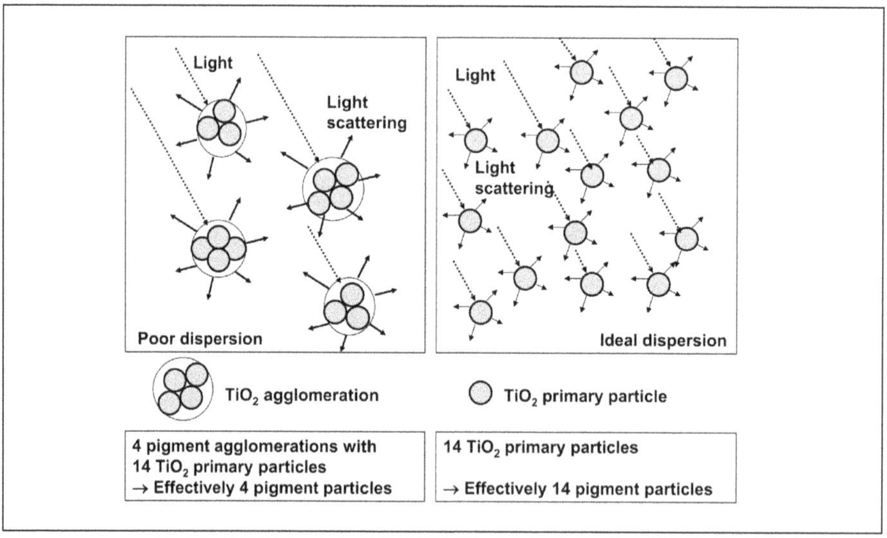

Figure 6.36: Schematic representation of optical efficiency with pigment primary particles and agglomerations in the case of titanium dioxide

The emulsion paint comparison described previously is impracticable with fillers whose oil absorption exceeds that of kaolin F: even at a level of 5 to 10 %, such fillers produce a massive increase in viscosity. Precipitated aluminium silicate, pyrogenous and precipitated silicic acids all belong in this category. As mentioned several times previously, products like these are used only as secondary fillers to boost opacity, as partial substitutes for white pigments, or to improve brightness or rheology. They are added at concentrations no greater than 5 % as a rule. A great many examples can be found in the producers' literature [7, 10]. The situation is similar with polymer hollow spheres: they too are deployed in emulsion paints as secondary fillers only, to boost opacity or as partial substitutes for white pigments.

Partial substitution of expensive white pigments like titanium dioxide works by artificially increasing the filler refractive index, see kaolin E and F as well as MCC A, or keeping the pigment particles apart by spacing. For spacing to work efficiently, the filler particles need to be of similar size to the pigment particles themselves, in other words 0.5 µm or less [11]. Long-lasting separation of pigment particles by fillers prevents the pigments from forming agglomerations; these are not unlike filler agglomerations, and equally undesirable. The reason is simple: a pigment agglomeration behaves optically as if it were a single particle. So achieving the same opacity in the presence of agglomerations necessitates more pigment than if the pigment primary particles were properly dispersed and stabi-

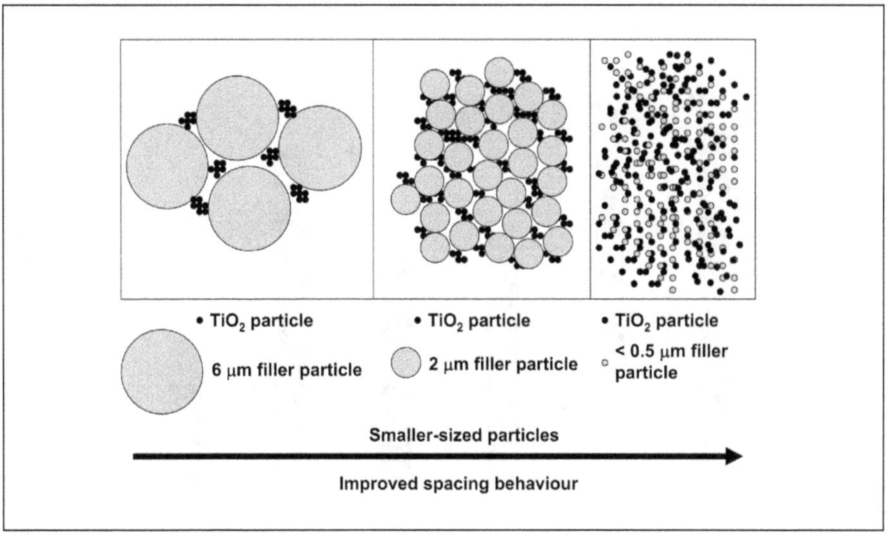

Figure 6.37: Schematic representation of the separation effect with titanium dioxide, and its efficiency as a function of filler particle size

lised. Eliminating pigment agglomerations is simply a means to improve the optical efficiency of the coating system, see Figure 6.36.

Filler can positively contribute to deagglomeration and stabilisation of pigments like titanium dioxide. This functionality improves with the fineness of the filler particles, because there is no space between them for the pigment particles to occupy. With large filler particles, on the other hand, there is sufficient space for pigment agglomerations in the packing voids between them. Figure 6.37 is a schematic representation of the separation, or distance effect (spacing) with fillers of differing fineness.

6.4.2 Colour properties

Although expectations of colour properties in fillers are definitely more modest than for pigments, requirements can be demanding all the same. These mainly come down to few key points:

- High whiteness/brightness
- Low yellowness index → neutral colour tone
- Low impact on colour strength in pigmented coatings.

Users select fillers especially for their whiteness. There can be large differences as already discussed in Chapters 4.1.1.1 and 5. Although varying degrees of whiteness tend

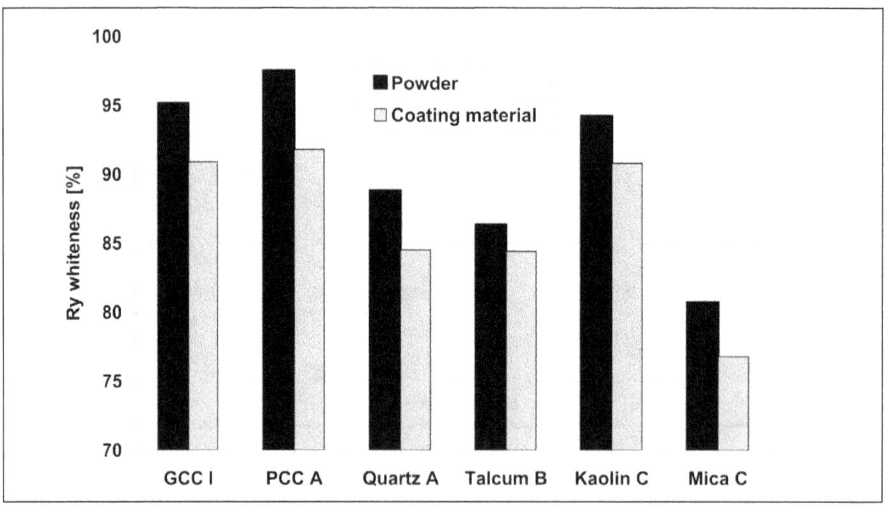

Figure 6.38: R_y whiteness comparison of various fillers, measured in powder form and in pigment-free emulsion paint with 70 % PVC

to be less pronounced in actual use, it is still true that the brightest fillers exhibit the highest whiteness in practice, and this gives them universal applicability. Less bright fillers, on the other hand, are automatically limited in their field of application and may indeed prove suitable only for primers and putties. Figure 6.38 shows a comparison of fillers with varying brightness, measured both as R_y whiteness in powder form, and after incorporation into the emulsion paint formulation of Chapter 6.4.1 with a PVC of 70 %. Precipitated and natural carbonates (GCC in this example) achieve very high whiteness, followed by calcined kaolins. The other layer silicates and quartz normally exhibit lower whiteness.

Low yellowness index makes it possible to formulate neutral white shades, and tint neutral and brilliant hues. Yellowness index, see also Chapter 4.1.1.3, is calculated from R_x, R_y and R_z colour values. It is the best means of expressing the visually perceived "yellow hue" in white colours. A positive side-effect of fillers with a high yellowness index is slightly increased opacity, because the whiteness over a white substrate is likewise not as high as with neutral brilliant white fillers. Figure 6.39 shows the yellowness index of various fillers in a similar manner to Figure 6.38, with the yellowness index of the powder by itself, and after incorporation into an emulsion paint.

Precipitated and natural carbonates exhibit the best values for yellowness index, just as they do for R_y whiteness. The further rankings are analogous to the R_y whiteness results shown previously.

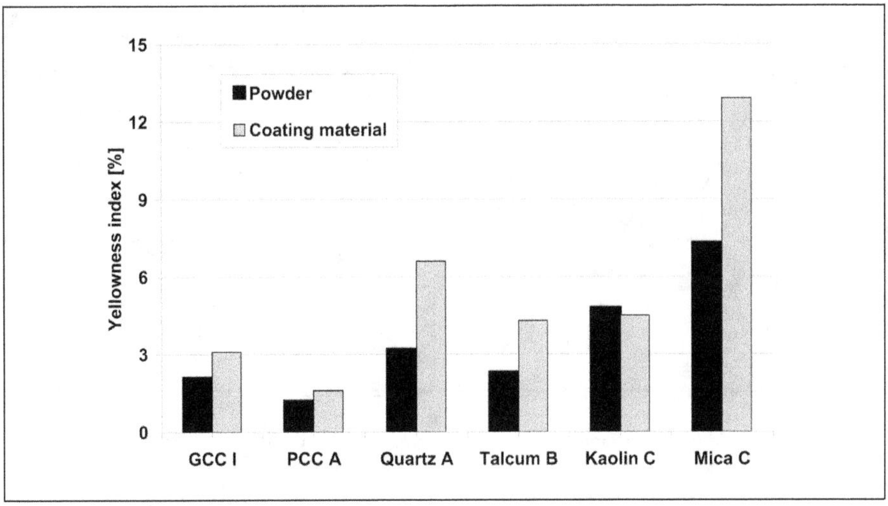

Figure 6.39: Yellowness index comparison of various fillers, measured in powder form and in pigment-free emulsion paint with 70 % PVC

In bright-hued formulations, the filler should ideally have no or at most minimal influence on colour strength or depth. But this is never the case in practice. Fillers absorb binder and thus reduce the residual amount available, which in turn reduces the colour depth. Accordingly, coarse fillers and minerals with very low oil absorption make the best choice for coating materials where pigment provides colour – a fact confirmed by weathering tests on exterior systems like façade paints. Differences of colour tone in excess of 5 dE can result from formulating full-tone paints with fillers of greatly differing oil absorption. It would be meaningless to present such differences in a greyscale illustration, so readers should instead refer to the corresponding literature [12, 13].

6.4.3 Reflectivity

In addition to colour tone and hiding power, optical reflectivity is also an important factor in the appearance of painted surfaces. Intensity of reflection can produce a range of subjective impressions: the human eye perceives strongly directional reflectivity as high gloss, for example. Gloss for its part suggests value, cleanliness, hygiene, as well as certain coldness. Surfaces that reflect light diffusely make a matt impression. Subjectively, a matt finish conveys a pleasant, warm mood. Impressions like these figure large in the way surfaces of objects are coated. Vehicles are a prime example: an automobile must have an ultra-high gloss finish in order to be considered new and valuable. One that has matt paintwork is automatically perceived as ancient and of little value. For residential interi-

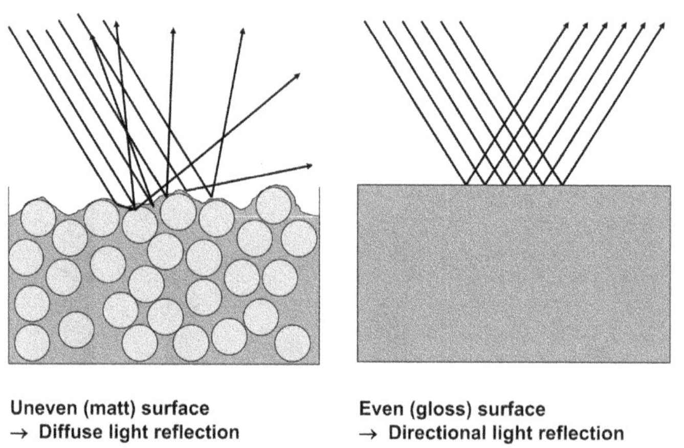

Figure 6.40: Schematic representation of directional reflection (gloss surface) and diffuse reflection (matt surface)

ors, though, matt is the finish of choice for walls and ceilings: people regard matt as more pleasing and cosy than shiny gloss, which suggests a sterile environment that is far from relaxing or welcoming. Fillers can massively influence the reflectivity of coating surfaces.

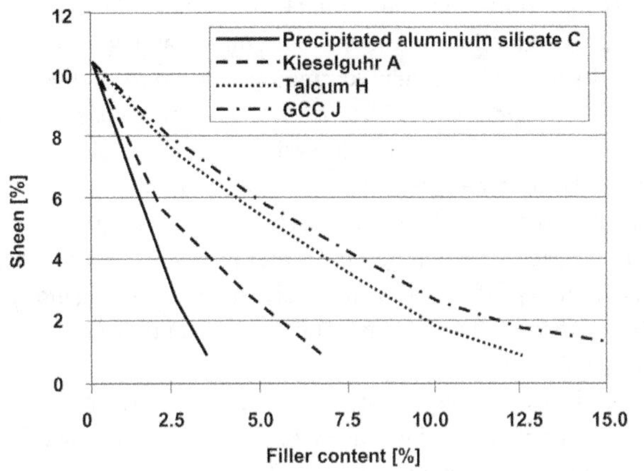

Figure 6.41: Matting material content of various fillers/matting agents and their effect on sheen in interior emulsion paint

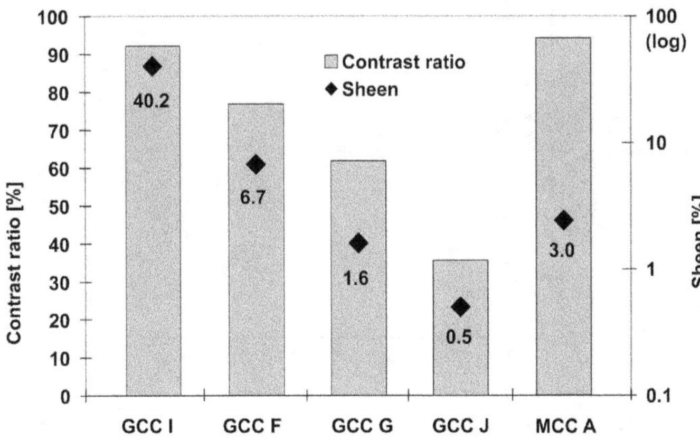

Figure 6.42: Effect of particle size on sheen and contrast ratio in a test emulsion paint

To appreciate this, it is necessary to understand just what makes a matt or a gloss surface. Figure 6.40 represents them schematically.

Surfaces are characterised by detecting the amount of reflected light at identical angles of incidence and reflection, and multiplying the ratio of reflected to incident intensity by 100 %. Total light reflection would thus result in a figure of 100 %, while the figure for extremely diffuse light reflection ("dead matt") would be under 1 %. In the paints and coatings industry, surface gloss or mattness is determined at angles of 20°, 60° or 85° to the normal. Reflectivity at an 85° angle is commonly referred to as sheen.

Emulsion paint formulations with a PVC of 50 % and above generally use coarse GCC, talcum or mica fillers to produce a matt finish. Such fillers make a very uneven coating film that reflects diffusely. The degree of mattness is regulated by the quantity and fineness of filler used. Coarse fillers with very high oil absorption, like kieselguhr and diatomaceous earth, precipitated silicic acids, or aluminium silicate can just as well be used. Indeed, their higher binder consumption produces a considerably superior matting effect [14–18]. On average, only one-half to one-third the quantity is needed in comparison to coarse fillers with low oil absorption, see Figure 6.41.

Beyond the quantity of filler used, particle size also has a strong influence on sheen. Depending on the amount of coarse fillers used for matting, hiding power can be impaired unless compensated by white pigment. Figure 6.42 uses various grades of GCC plus MCC A to illustrate the effect of filler fineness on sheen in a test emulsion paint (similar composition to that in Chapters 6.4.1 and 6.4.2) [19]. Despite a fine particle size distribution, the before described MCC A, which has a strong positive effect onto opacity, shows low reflectivity. This matting effect can be explained by the very amorphous surface and high oil absorption of MCC A.

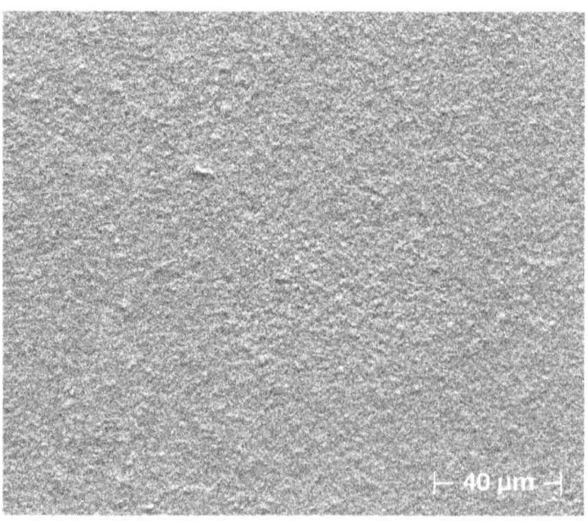

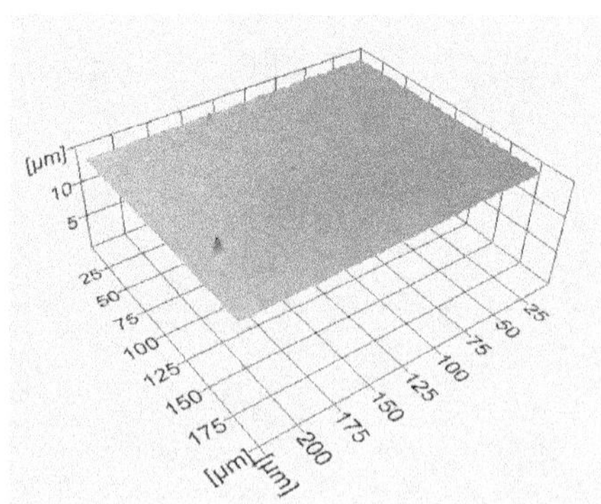

Figures 6.43 through 6.44: Surface of satin matt emulsion paint as an SEM image (6.43) and CLSM profile (6.44)
source: Omya AG

The schematic representation of Figure 6.40 actually has quite a similar appearance in practice. Scanning electron microscopy (SEM) provides one means of characterising coating film surfaces. Confocal laser scanning microscopy (CLSM) is another technique that delivers a greatly superior spatial view of the coating surface structure. Figure 6.43 shows a SEM micrograph of a satin matt emulsion paint film. Figure 6.44 shows the same paint as a three-dimensional CLSM profile. The very smooth coating film surface is plainly apparent.

Contrasted with the smooth surface of satin matt emulsion paint, the film surface of matt emulsion paint is quite spectacular. Figure 6.45 shows a SEM micrograph of a matt emulsion paint containing 20 µm filler particles. It has a correspondingly lively CLSM profile, see Figure 6.46.

Although they are good for matting, fillers are not especially useful in high-gloss coating systems. In fact, these systems (gloss at 20° >80 % and at 60° >90 %) use pigments alone. Inasmuch as fillers find use in gloss paints and coatings, the quantities involved are very small, on the order of 0 to 8 %. Their function here is not to enhance the gloss, but to space the pigment particles and prevent optically inefficient agglomerations.

6.4.4 Mechanical properties

There are numerous tests for characterising the mechanical properties of coating films. The type and extent of tests used depends on the application area.

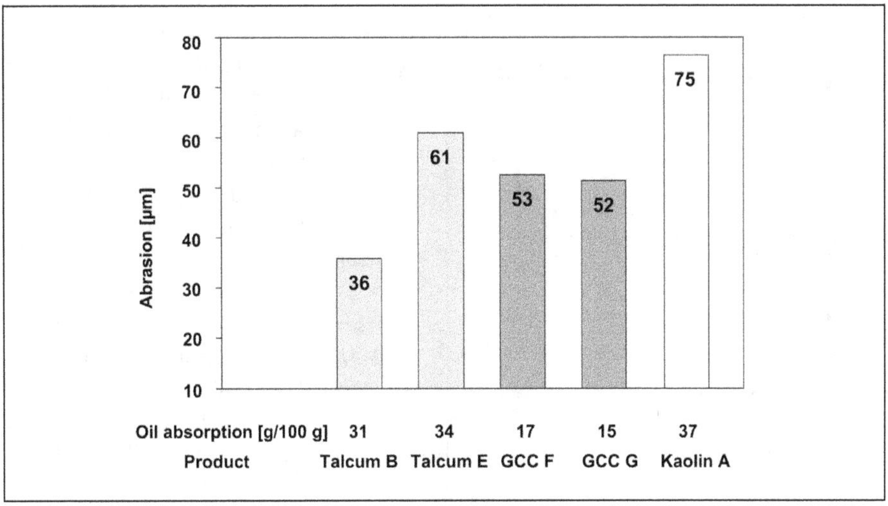

Figure 6.47: Wet scrub resistance according to EN 11998 for interior emulsion paints formulated using fillers with different binder consumption

Tests on emulsion paints for decorative coatings are largely limited to determining wet scrub resistance according to EN 11998. The depth of scrubbing produced by a scouring pad is measured in micrometres after a defined number of double strokes (200 or 40, depending on quality and classification). Wet scrub resistance is related to the binder consumption of the filler used. At a similar PVC, fillers with higher oil absorption exhibit greater porosity in emulsion paints than fillers with low oil absorption. It follows that in coating materials containing fillers with high oil absorption, the bonds between pigments, fillers and binder are less stabilised. In consequence, fillers with low oil absorption exhibit greater wet scrub resistance at given PVC than do highly absorbent fillers, see Figure 6.47.

The effect described above – fillers with low oil absorption exhibit less scrubbing – holds true only in part. In Figure 6.47, talcum exhibits superior wet scrub resistance to both GCC grades with lower oil absorption. This is evidence that wet scrub resistance is influenced not only by oil absorption, but other filler properties as well. Talcum's superior wet scrub resistance stems from its lamellar particle structure, hydrophobicity and the high aspect ratio of the talcum platelets. Natural kaolin exhibited most scrubbing in this example.

Shrinkage crack-free drying of thick coating layers is very important in decorative coatings like emulsion paints, and especially in façade paints that are sometimes applied to very rough and relatively uneven substrates (renders and plasters). Depending on the substrate, coating thickness may be normal or considerably increased as a result. Large drying cracks or other flaws allow water to penetrate the coating, render or plaster and damage the underlying masonry. Shrinkage crack-free drying is tested by applying coating material to a wedge with a gap that progressively deepens from 0 mm to 2 or 3 mm maximum, and letting it dry. The paint is examined for cracks after a sufficient interval, typically around 3 days at 23 °C and 50 % relative humidity. In Figure 6.48, the filler packing grows coarser from top to bottom: 20 parts fine filler (top) is progressively exchanged for coarser filler in 2-part increments, down to 10 parts each of fine and coarse filler (bottom). The result is clearly recognisable and measurable. In the topmost wedge, the crack-free dry film thickness is only 500 µm or so, while the bottommost wedge reveals a crack-free dry film thickness in excess of 2000 µm.

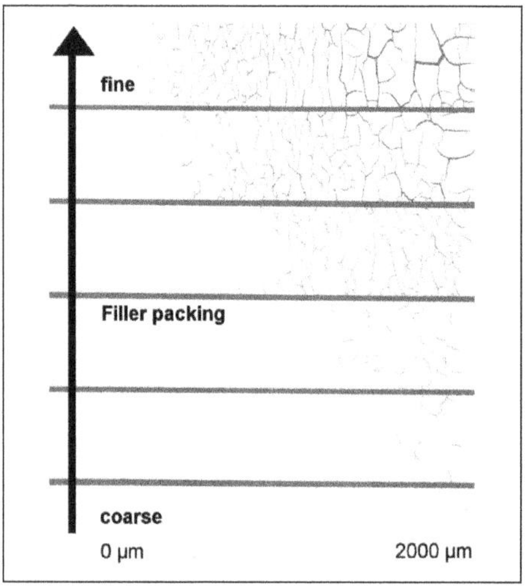

Figure 6.48: Crack formation in coating films with differing filler packing in a 2 mm wedge
source: Kronos International, Inc.

Industrial coating materials call for a different set of mechanical properties, which is why other tests are also used. Some typical tests include:
- Adhesion strength (ISO 2409) also known as cross-hatch test
- Erichsen indentation (ISO 1520)
- König pendulum hardness (ISO 1522)
- Impact test (ISO 6272)
- Mandrel bend test (ISO 1519).

The list is by no means complete, but provides an impression of the requirements and tests involved. Fillers in industrial coatings can influence mechanical properties like hardness, flexibility and adhesion. However, these properties primarily stem from the binder, or a combination of binders and their film hardening by various chemical reactions. So, although fillers doubtless play a supporting role in a coating's mechanical properties, they are not the main factor.

Quartz and other fillers with high Mohs hardness are ideal in situations that call for exceptional scrub resistance. In fact, this is why they are frequently used in heavy-duty industrial floor coatings. Talcum, on the other hand, is a filler that improves adhesion properties. Moreover, talcum has very low Mohs hardness, which is why it is often found in putties. Barium sulphate and calcium carbonate fillers exhibit only modest binder consumption, so they are deployed a great deal as universal fillers in industrial coatings, where they contribute to a generally well-balanced properties profile.

6.4.5 Chemical resistance

Industrial coating systems, automotive coatings and anti-corrosion systems have to meet rigorous performance criteria. Next to aesthetic and mechanical requirements, they must also stand up well to chemicals like automotive fuel, cleaning and care products, as well as corrosive compounds like acids and alkalis. There are various chemical resistance tests in use, of which the two most important by far are:
- Resistance to chemicals (ISO 2812)
- Salt spray test (ISO 7253).

ISO 2812 describes tests for chemical resistance to a very broad range of substances. Testing a dried or cured coating film involves exposing it to reagents like strong acids and bases, solvents, hydrocarbons, and other aggressive substances like mustard, lipstick, etc. for a defined interval like 1 or 24 hours. Following exposure to the chemicals, the coating surface is examined for evidence of discoloration, blistering, loss of adhesion, etc. It is worth noting here that resistance to chemicals is primarily a function of the binder system, because all of the coatings in question are under-critically formulated. Hence, the filler particles are protectively enveloped in binder. Yet for ultimate chemical resistance, there are still benefits to using chemically inert fillers: should the coating film be damaged, a chemically inert filler will not react. Examples of such fillers are: barium sulphate, layer silicates like talcum, kaolin, mica and feldspar, as well as quartz and cristobalite. Carbonate fillers as a group have no resistance to acids.

The salt spray test is very commonplace and far more equipment-intensive than chemical resistance testing. A 5 % salt solution is sprayed as a fine mist in a climatic chamber under standard conditions of 35 °C and 100 % air humidity. The spray lands on the surface of the coating under test, which was previously damaged in a defined manner, right through to the substrate. This enables ions and other corrosion-inducing agents to penetrate as deep as the substrate, which is frequently steel. From there, it can creep under the coating material and thus impair its resistance through poor adhesion. Blisters may subsequently form on the paint film, and flake off. Corrosion products form on the metallic substrate, causing a volume change and possibly a similar loss of adhesion. A coating material's protective properties largely come down to the binder system used. Anti-corrosion pigments like zinc phosphate can also help. Filler plays a role inasmuch as corrosion-inducing agents can penetrate the coating film and start reactions. It follows that chemically inert substances like barium sulphate, talcum and mica are the fillers of choice. Yet acid-soluble fillers like calcium carbonate do have a place in anti-corrosion systems [3]: carbonate fillers are basic and buffer a decreasing pH, which has a passivating effect on metallic substrates.

Chemical and corrosion-resistant coating systems are frequently formulated in a highly individual style, tailor-made for one specific application. It is extremely rare in practice to find an anti-corrosion application that uses just one filler; mixtures of various mineral fillers are frequently encountered. Blends of lamellar and nodular fillers are popular, with lamellar fillers like talcum and mica being favoured for their barrier properties. Lamellar fillers lengthen the diffusion path through the coating, thus preventing or delaying the onset of corrosion. Moreover, talcum increases the coating system's hydrophobicity and adhesion. Combining lamellar with nodular fillers achieves good filler packing. However, the fillers selected should not be too coarse, which would produce excessive microporosity and accelerate diffusion processes. Figure 6.49 shows an SEM cross-sectional micrograph of an anti-corrosion primer.

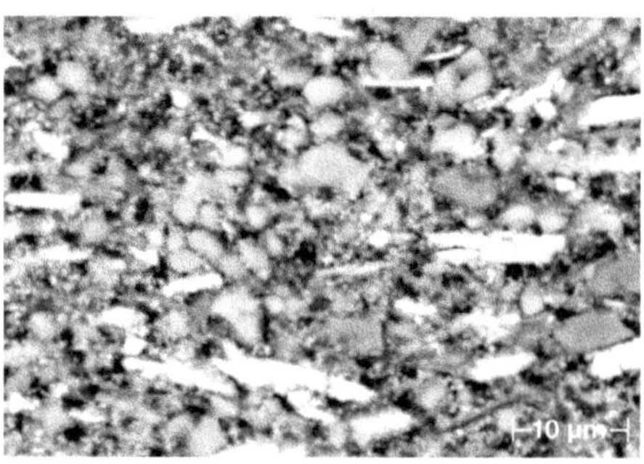

Figure 6.49: SEM cross sectional view through a water-based anti-corrosion primer, showing the packing density of nodular and lamellar fillers, plus pigments source: Omya AG

The nodular and lamellar fillers and anti-corrosion pigment are all clearly apparent. The corresponding SEM/EDS analysis image further helps to localise and identify the fillers present. The top left frame shows the overall image, top right is the image for magnesium silicate (talcum), bottom left shows the zinc phosphate anti-corrosion pigment, and particles of calcium carbonate are visible at bottom right.

6.4.6 Outdoor durability

Every coating material intended for exterior use has to be capable of withstanding the elements for a certain period, so that coated items retain their value and functionality. As with most forms of resistance, it is the binder that plays the main protective role. Fillers cannot compensate shortcomings here.

Weather resistance can be tested naturally, by outdoor weathering. The disadvantage here is the extended weathering period that is necessary for producing a meaningful result. For this reason, supplementary artificial weathering is performed with the help of accelerated weathering testers. Methods include accelerated weathering by a QU-V weathering tester (UV-A or UV-B irradiation), a Weather-O-Meter (xenon or carbon arc lamp), or a

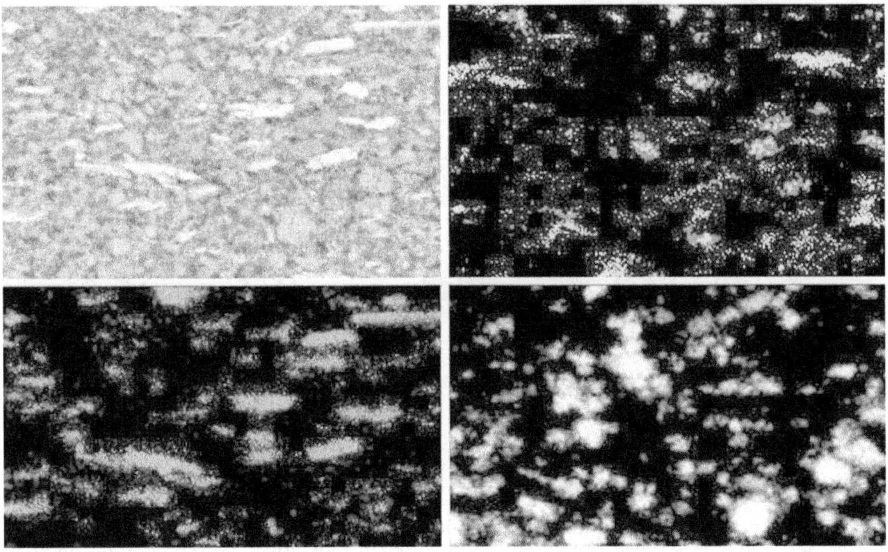

Figure 6.50: SEM/EDS analysis image, for identification and localisation of filler and pigment particles in a water-based anti-corrosion primer. overall image (top left), talcum particles (top right), zinc phosphate particles (bottom left), and calcium carbonate particles (bottom right)

source: Omva AG

Suntest (UV light). Although testing times are shorter in every case, correlation with outdoor weathering is slight at best. It is therefore essential to test a known standard alongside, in order to reference the results. Accelerated weathering tests may run for several

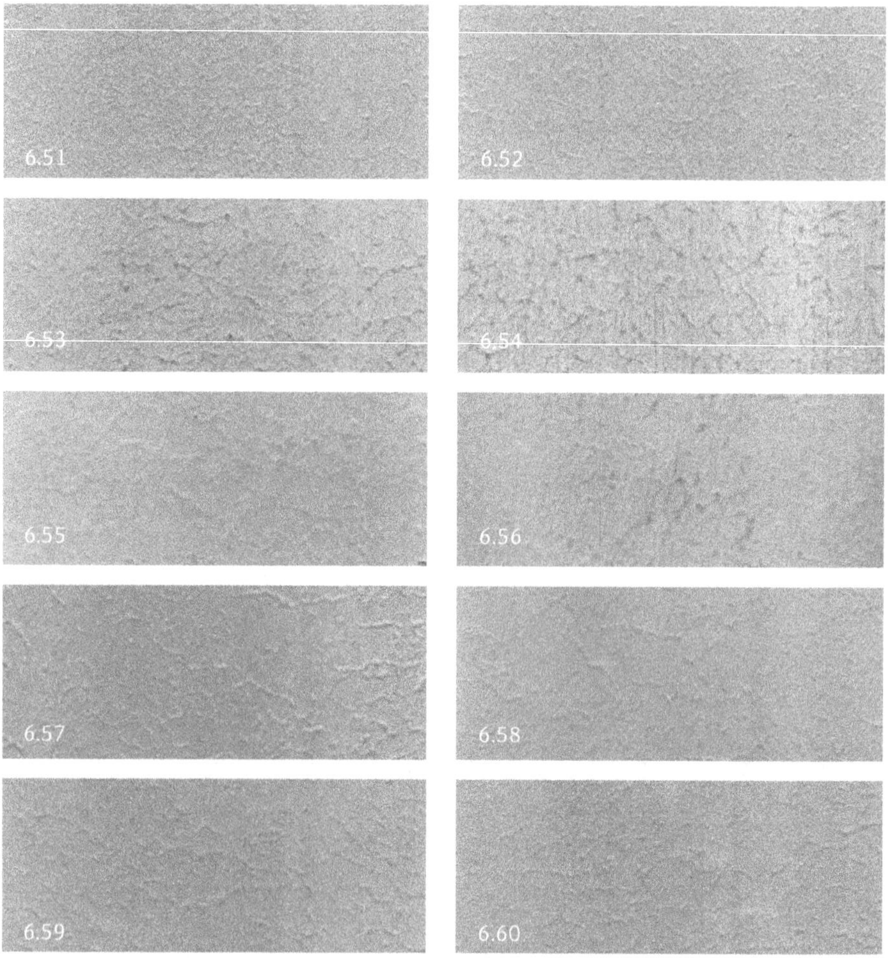

Figures 6.51 through 6.60: Blue pure acrylate full-tone façade paint with 42 % PVC unweathered, and after 2 years of outdoor weathering in Switzerland (400 m altitude, angled 45° to the south), GCC H unweathered (6.51) and weathered (6.52), kaolin D unweathered (6.53) and weathered (6.54), talcum C unweathered (6.55) and weathered (6.56), barium sulphate A unweathered (6.57) and weathered (6.58), cristobalite A unweathered (6.59) and weathered (6.60

thousand hours, depending on the application. The following properties are measured during and after the weathering test:
- Colour retention or difference
- Gloss retention
- Chalking.

Barium sulphate fillers are frequently used in exterior applications with low PVC, such as powder coatings. Precipitated barium sulphate in particular has excellent weathering resistance to UV radiation, with no tendency to chalking. Colour and gloss retention are similarly superior [20]. It all means that precipitated barium sulphate meets the highest demands of exterior coatings. It needs mentioning here that the term "chalking" has nothing to do with chalk, the mineral. In fact, it describes a weathering mechanism that causes binder at the film surface to degrade. The top layer of pigment and filler particles becomes exposed and unprotected; so simple hand contact will brush them away. With white coatings, the chalking is likewise white, so fingers and hands look like they have just wiped a school blackboard. In the Helmen chalking test, loose pigment and filler particles are lifted off the surface by a length of adhesive tape and visually compared against a scale of 1 to 10. 10 is the best value and corresponds to no chalking.

Large quantities of calcium carbonate go into emulsion paints, renders and plasters for façade coatings. Traditional local deposits of dolomite, kaolin, barium sulphate, cristobalite and feldspar are also used to some extent. However, not all fillers perform equally well. Field trials and studies of critical full colour emulsion paints have produced very good weathering results for silica-free fillers like calcium carbonate, dolomite and barium sulphate. High-silica fillers like mica, talcum, kaolin and quartz can occasionally exhibit considerable chalking [12, 13]. The outdoor weathering tests illustrated in Figure 6.51 through 6.60 were performed using wood panels coated with blue full-tone façade paint based on pure acrylate and colanyl blue at 42 % PVC. The figures compare the unweathered sample with the result after two years' outdoor weathering.

Fillers for exterior applications should also have low oil absorption, so that strong hues in particular are not excessively lightened. This avoids using more coloured pigment than necessary. The obtained colour changes have been summarized in Figure 6.61.

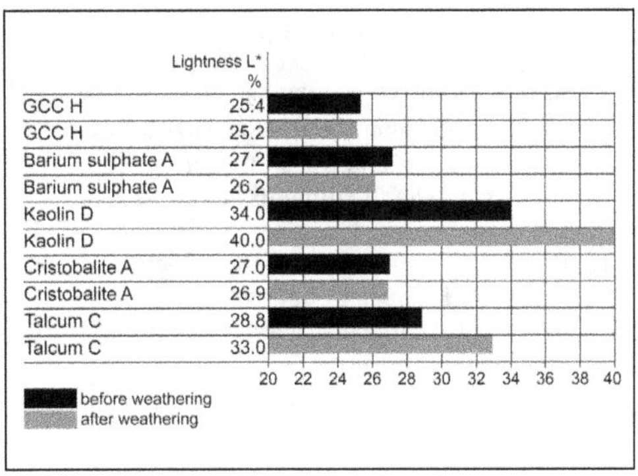

Figure 6.61: Initial colour readings of exposed panels and changes after weathering

6.5 References

[1] Grochal, P., Troll, M., "Fehler vermeiden", Jg. 107, S. 119, Farbe + Lack, Vincentz Verlag, Hannover, 9/2001

[2] Brock, T., Groteklaes, M., Mischke, P.; "European Coatings Handbook", 2nd edition, S. 193–199, Vincentz Verlag, Hannover, 1998

[3] Gysau, D., "Influence of Extenders and Extender-Combinations on the Anti-Corrosion Properties of a Water-Thinnable Primer", XXIV. Fatipec Congress, Interlaken, 1996

[4] Dörr, H., Holzinger, F., "KRONOS Titandioxid in Dispersionsfarben", Kronos Titan-GmbH, Leverkusen, 1989

[5] Gysau, D., Merz, M., Chetelat, E., "OMYACOAT 250- und 850-OG – zwei neue Füllstoffe für Farben und Lacke", p. 18–24, Welt der Farben, Köln, 4/2003

[6] Carlozzo, B.J., "The Use of Spherical Extenders in VOC Compliant Coatings", Jg. 69, p. 71–84, Journal of Coatings Technology, 7/1997

[7] Degussa documentation, "Hydrophobe Aerosil®-Typen und ihr Einsatz in der Lackindustrie", Schriftenreihe Pigmente No 18, Degussa, Frankfurt

[8] Coatex documentation, "Dispersants and Applied Rheology", Training documentation, Coatex S.A., Genay. 2003

[[9] Tego documentation, "tego journal", 2nd edition, p. 77–84, Tego Chemie Service GmbH, Essen, 2002

[10] Degussa Documentation, "Sipernat 820 A for Emulsion and Decorative Paints", Technical Bulletin Pigments No. 34, Degussa, Frankfurt

[11] Dietz, P. F., "Spacing for better effects", p. 14–20, European Coatings Journal (ECJ), Vincentz Verlag, Hannover, 7–8/2003

[12] Omya documentation, "Einfluss mineralischer Füllstoffe auf die Wetterbeständigkeit und Farbtonhaltung von bunten Fassadenfarben auf Reinacrylat Basis", ATS Bulletin No 165, Omya, Oftringen, 2004

[13] Schwartz, M., Kossmann, H., "Acrylic and acrylic/styrene copolymer dispersions", p. 134–140, European Coatings Journal (ECJ), Vincentz Verlag, Hannover, 3/1998

[14] Omya documentation, "CALCIMATT – Ein neues Mattierungsmittel für Innendispersionsfarben", ATS Bulletin No 202, Omya, Oftringen, 2004

[15] Omya documentation, "Vergleich der Mattierungsmittel CALCIMATT und Diatomeenerde in matten ELF-Innendispersionsfarben", ATS Bulletin No 205, Omya, Oftringen, 2004

[16] Degussa documentation, "Acematt® - Mattierungsmittel für die Lackindustrie", Paper Fine Particles No 21, Degussa, Frankfurt

[17] Katholnig, P., "Neue umweltverträgliche Wege der Mattierung", VILF paper No 5, 2003

[18] World Minerals Documentation, "Optimat performances vs. Diatomaceous Earth", Technical Bulletin Optimat, World Minerals, Nanterre

[19] Lehner, F., "Calcium Carbonate in Paints and Coatings", 2nd International Paint and Varnish Conference Ukraine, Crimea, 2004

[20] Sachtleben documentation, "Sachtleben Blanc fixe Product Information – Benefits in Powder Coatings", Sachtleben Chemie, Duisburg

7 Trends

The production of mineral fillers on an industrial scale dates back over a century at the time of writing. Filler products have undergone much change during this period, as have their applications. Demand for fillers of increasing quality in terms of brightness, purity, functionality and product consistency drove changes to existing processes, as well as development of new production techniques. Grinding processes for natural minerals have likewise tracked steadily increasing demand. Where dry grinding once predominated, wet grinding processes have made considerable inroads. Wet grinding is a rather more efficient means of producing the very fine fillers for which demand has been growing recently, and it increases the options available to developers of new filler materials. Meanwhile, there are some relatively new technical processes for imparting greater functionality to fillers. Developments in this area include producing fillers by chemical precipitation reactions, along with modification of natural fillers by thermal treatment, or coating them with other materials [1].

Nanotechnology is the term most often heard in association with contemporary innovations in paints and coatings, and raw materials. Yet it is frequently used as a USP for products and materials that do not yet truly deserve the designation. There are very few genuine nano-products at present in the filler arena, although the number looks set to rise in future. Nanotechnology holds out the promise of products with enhanced properties, and hence greater functionality.

Next to demand for products with greater functionality and the associated development effort, there is also evident demand for fillers that increase the production efficiency of paints and coatings. This is just as true for fillers as it is for pigments. Developments targeted at greater filler efficiency currently focus on simpler handling, both for getting the material where it is needed, and incorporating it into coating systems.

Recent trend in the mineral filler industry as well as in industries consuming mineral fillers is the topic sustainability. Of course, this was considered already in the past, but obviously less intensive as demanded now from public and other stakeholders. Sustainability is a wide field, where in particular the buzzword "Carbon Footprint" and Life Cycle Assessment (LCA) dominating.

Table 7.1: Commercial fillers with median particle sizes in the nanometre range

Filler	Median particle size [nm]	Agglomeration size [µm]
Precipitated silicic acids	5 to 100	–
Pyrogenous silic acids	5 to 50	1 to 40
Silica gels	3 to 20	1 to 20
Precipitated calcium carbonate	30 to 350	–

Light weight fillers (LWF) become popular as well. The true LWF's are hollow microspheres and can be based on different chemical composition, e.g. glass, ceramic, fly ash as well as polymer based. They provide a lot of benefits besides the specific weight reduction of formulations.

7.1 Nanotechnology

The filler industry was a very early mover in nanomaterials and their production, so nano is a pervasive buzzword here as well. But ahead of any discussion about nano-products and their process possibilities, some clear definitions are in order. After all, not every product that has been embellished with a nano prefix really merits the term. In theory, an ultra-fine ground natural filler with a D50% median particle size of 0.8 µm, or 800 nm, might be described as a nano-filler. Yet such a product does not match the consensus view of nano-products. So, just what are nano-products and how are they defined? Regrettably, there is still a lack of consistent terminology here. From observation of discussions and publications, though, it is nonetheless reasonable to conclude that materials with a particle size below 200 nm, and certainly those below 100 nm, can safely be termed nano-products [2].

The International Organization for Standardization (ISO) has been working on developing vocabulary and core terms for nanomaterials and nanotechnologies since 2005; the resulting vocabulary and core terms refer to the size range between approximately 1 nm and 100 nm [3]. The responsible European Commission drafted the following recommendation. Nanomaterial means a material that meets at least one of the following criteria:

- consists of particles, with one or more external dimensions in the size range 1 nm – 100 nm for more than 1 % of their number size distribution
- has internal or surface structures in one or more dimensions in the size range 1 nm – 100 nm
- has a specific surface area by volume greater than 60 m^2/cm^3, excluding materials consisting of particles with a size lower than 1 nm.

As mentioned at the outset, the filler industry was an early developer of nano-prod-ucts. Table 7.1 lists some examples of commercial products that have found applications in the market for years, if not decades. Currently available nano-fillers produce a variety of technical effects in coating material applications:
- Rheology
- Stability
- Reinforcement
- Matting
- Dispersion
- Adhesion strength
- Anti-corrosion properties.

Beyond such existing offerings, further manufacturing developments are underway in nano-fillers, where nanotechnology is used for imparting specific properties to a variety of fillers and mineral stocks. It is quite feasible for new nano-fillers like precipitated silicic acids and silica gels to result in end products not in the nanometre, but rather the micrometre range [4]. Such results would greatly simplify nano-product stabilisation as well as handling. Contemporary development work aims to supplement existing nano-filler offerings and further improve their technical properties in applications. The possibilities include:
- Greater mechanical strength, e.g. scratch resistance
- Less soiling of coating films
- Easy cleanability
- High absorptive capacity
- Enhanced light scattering for greater opacity
- Better heat resistance.

As well as technical progress, environmental and safety aspects must not be neglected. The particle size of nano-products not only makes their surfaces very reactive, it also allows them to penetrate cell membranes. Depending on the size of the particles and their chemical composition, it is conceivable that nano-products could have negative influences on health and the environment [5, 6].

Compared to conventional fillers, nano-fillers are considerably more complex to develop and implement as manufacturable products. Realisation will thus be more cost and time-intensive than processes and products developed by the filler industry to date. Nevertheless, we can look forward to new and interesting fillers with some very special properties.

7.2 Forms of delivery

Progressive rationalisation by industry at large, including the paints and coatings sector, is driving changes in development, testing, production, and usage. Especially in the production of coating materials, fillers bear strongly on:
- Dispersibility
- Incorporation
- Handling.

All three have a common denominator, namely that they are recurring issues with fillers in very fine powder form. Moreover, fillers with a very low bulk density are especially disadvantageous for incorporation in more viscous systems, due to the levels of dust that develop and contaminate the surroundings where manual processes are involved, as well as the difficulties of handling such material in larger quantities. Such problems can be ameliorated, and sometimes remedied altogether by the following forms of delivery:
- Compacted fillers with a higher bulk density and better flowability and moveability
- Granulated or agglomerated fillers for better handling in larger quantities
- Fluidised fillers in aqueous suspension, known as slurries. These are easily handled and incorporated in large volumes without dispersion[7], but they are only relevant to water-based coating materials.

All the forms of delivery mentioned have their advantages, as well as certain limitations and disadvantages. Various products in the forms listed above are currently in the trial phase, or are already in use by a few early adopters among coating material producers. Some of these technologies and developments are still in their infancy, and show further progress and optimisation potential. Here too, the future looks exciting.

7.3 Sustainability

Some of the most frequent buzzwords in Sustainability are "Carbon Footprint" and "Life Cycle Assessment" (LCA). The carbon footprint is a subset of the more comprehensive LCA and defined as "the total set of greenhouse gas (GHG) emissions caused by an organization, event, product or person".

The carbon footprint is measured in tonnes of carbon dioxide equivalent (tCO_2e). The carbon dioxide equivalent (CO_2e) allows the different greenhouse gases to be compared on a like-for-like basis relative to one unit of CO_2. The footprint considers all six of the Kyoto Protocol greenhouse gases: Carbon dioxide (CO_2), Methane (CH_4), Nitrous oxide (N_2O), Hydrofluorocarbons (HFCs), Perfluorocarbons (PFCs) and Sulphur hexafluoride

(SF_6). CO_2e is calculated by multiplying the emissions of each of the six greenhouse gases by its 100 year global warming potential (GWP).

Life cycle assessment is increasingly recognised as one of the best tools for assessing the overall impact of a product – or service - as it covers every consumption and emission related to the whole life cycle of the product, i.e. its production, use and end of use. According to ISO 14040 LCA is defined by as the "compilation and evaluation of the inputs and outputs and the potential environmental impacts of a product system throughout its life cycle". LCA is referred to in a growing number of environmental and product based regulations and policies and will no doubt become an influential tool in the decision making process related to the development and implementation of these policies. To encourage and facilitate the use of this tool, in 2003, the European Commission issued a Communication on "Integrated Product Policy: Building on Environmental Life-Cycle Thinking" in which it called for the development of a platform to facilitate communication and exchange of LCA databases.

Following this Communication a European Platform on Life Cycle Assessment (EPLCA) has been created which aims – amongst others – to develop a European reference Life Cycle Data System (ELCD). To complete the database, European associations are invited to collaborate on its development by providing them with accurate Life Cycle Inventory (LCI) data. Data provided by business associations would then be considered as the reference LCI data. The European calcium carbonate and dolomite producers (CCA-Europe) have integrated the latest results in the ELCD, published on the Joint Research Centre of the European Commission [8].

Recognising the importance of such a database, IMA-Europe actively participated to the project by developing LCA data on the following major products of its members for paint applications:

- Calcium carbonate (<63 μm)
- Micronised talc
- Dry silica sand
- Coarse kaolin

The LCI datasheets are ISO 14040/14044 compliant and were developed by Ecobilan, an independent consulting firm which is renowned internationally for its expertise on LCA. The data contained in the abovementioned datasheets are cradle to gate data and therefore do not cover the use and end of use of the minerals nor their transport to downstream users. The LCI data were developed using generic process data - i.e. quarrying, crushing, drying, grinding, packaging, etc- as production steps. Consumption and emission data were collected for each of the production steps from all the participating sites and aggregated as averages. The final product based LCI data were obtained by adding the appropriate production steps related data [9].

Mineral fillers, in particular natural ones, do have low carbon footprints compared with other raw materials for paint formulations, e.g. dry ground calcium carbonate 62 kg CO_2/ton filler[10]. Therefore, mineral fillers can add value in form of lowering the total carbon footprint of paint formulations. This is particularly interesting when benchmarking decorative paints from different geographical regions. By tradition and related to historic filler supply sources, decorative paints in UK, US and Asia-Pacific are formulated at much lower PV, because of using mineral fillers with much higher oil absorption. On continental Europe these kinds of paints are dominantly formulated with mineral fillers having much lower oil absorption. Thus the demand of polymer to achieve similar wet scrub resistance is much lower. In terms of carbon footprint, decorative paints on Continental Europe having statistically a lower carbon footprint compared with decorative paints in the US, UK and Asia-Pacific.

7.4 Light weight fillers

Typically light weight fillers are hollow microspheres in the scale of micrometers. They can be manufactured form different materials such as glass, ceramic, fly ash (cenospheres), polymers and minerals. During the manufacturing process they obtain their hollow structure, e.g. rapid thermal treatment at very high temperatures, often at more than 1000 °C, expands the previously solid micro particles to hollows. Based on the structural properties light weight fillers can support the following properties:

- Low ratio of shrinking and deformation
- Less cracking when applied at higher thickness
- Reduced sagging
- Lower specific weight of final formulations and thus logistic costs
- Easier and smoother application of coatings

Thus various applications and markets such as construction composites, paints & coatings, plasters, renders, joint fillers and sparkling compounds rely on light weight fillers. The industry demand is steadily increasing and the compound annual growth rates are above 10 %[11]. The LWF market is dominated by glass microspheres which have a share of approximately 40 %, followed by microspheres based on ceramic, fly ash and polymer. Mineral based light weight fillers are on a rather low volume, but providing more opportunities in future.

7.5 References

[1] Gysau, D., "Extender or solution provider", p. 50, European Coatings Journal (ECJ), Insider Edition, Vincentz Network, Hannover, 1/2005

[2] EC Communication of 12 May 2004: "Towards a European Strategy for Nanotechnology", COM 388 final, 2004

[3] ISO Concept Database "Vocabulary and core terms for nanomaterials and nanotechnologies", ISO/TC 229, http://cdb.iso.org, 2010

[4] Gysau, D., "Rosen und Golfbälle", vol. 107, Farbe + Lack, Vincentz Verlag, Hannover, 9/2001

[5] Nanotechnologies: A preliminary risk analysis on the basis of a workshop organized in Brussels on 1-2 March 2004 by the Health and Consumer Protection Directorate General of the European Commission, Internet http://europa.eu.int/comm/health/ph_risk/events_risk_en.htm, 2004

[6] UK Royal Society and the Royal Academy of Engineering, "Nanoscience and nanotechnologies: opportunities and uncertainties", http://nanotec.org.uk/finalReport.htm, 2004

[7] Gysau, D., "Mischen statt Blechen", vol. 110, Farbe + Lack, Vincentz Network, Hannover, 12/2004

[8] European Life Cycle Database (ELCD), Joint Research Centre, http://elcd,jrc.ec.europa.eu/ELCD3/processlist.xhtml, 2008

[9] Industrial Mineral Association – Europe, "European Life Cycle Database Project – Methodology Report", Ecobilan, Neuilly sur Seine Cedex, http://www.ima-eu.org/fileadmin/downloads/sustainable/LCA_project/IMA__ELCD_Methodology_report_04.pdf, 2008

[10] Industrial Mineral Association – Europe, http://www.ima-eu.org, Brussels, 2010

[11] Markets and Markets, "Microspheres Market – Global Trends & Forecast to 2021", 2016

Examples for guide formulations

Guide formulation for satin gloss emulsion paint

Component	Parts by weight
De-ionised water	175.7
Tylose MH 15 000 YG8 [1] (thickener)	2.5
Calgon N [2], 10 % (wetting and dispersing agent)	12.5
Lopon 894 [2] (wetting and dispersing agent)	3.5
Agitan 315 [3] (defoamer)	4.5
Mergal K 9 N [4] (preservative)	2.0
Tafigel PUR 50 [3] (thickener)	2.3
Sodium hydroxide solution. 10 % (pH regulator)	2.0
Tiona RCL-535 [5] (white pigment)	230.0
Omyacoat 420 – OG [6] (filler)	85.0
Mowilith LDM 1871 [7] approx. 53 % (vinyl acetate/ethylene emulsion)	480.0
Total	**1000.0**

[1] Shin Etsu [2] BK Guilini [3] Münzing Chemie [4] Troy Chemie [5] Cristal [6] Omya [7] Celanese

Detlef Gysau: Fillers for Paints

Guide formulation for 1-component anti-corrosion primer

Component	Parts by weight
Resydrol VWE 6050 [1]. 40 % (epoxy-acrylic hybrid binder)	346.8
Acronal S 760 [2]. 50 % (butyl acrylic/styrene emulsion)	92.5
DMAMP-80 [3] (pH regulator)	1.6
Corrosion inhibitor L1 [4] (corrosion inhibitor)	0.3
Additol VXW 4940 [1] (drier)	4.2
Additol XL 270 [1] (anti-settling agent)	3.5
Additol XL 297 [1] (anti-skinning agent)	2.0
Additol XW 376 [1] (defoamer)	5.0
Bayferrox 130 M [5] (red pigment)	65.0
SICOR ZNP/M [2] (anti-corrosion pigment)	75.0
Micro-Talc A.T. Extra [6] (filler)	75.0
Millicarb-OG [7] (filler)	141.0
Rheolate 278 [8] (thickener)	6.5
Butyl diglycol (solvent)	2.0
Water	179.6
Total	**1000.0**

[1] UCB [2] BASF [3] Angus [4] Erbslöh [5] Lanxess [6] Mondo Minerals [7] Omya
[8] NL Industrial Chemicals

Guide formulation for 2-component epoxy primer

Component A	Parts by weight
Araldite GZ 7071 X 75 [1] (solid epoxy resin. based on Bisphenol A)	20.0
Araldite GY 783 [1] (reactive thinner, Bisphenol A/F)	15.0
Disperbyk 110 [2] (dispersing agent)	1.0
Troythix 150 ACS [3] (thickener)	3.0
Heucophos ZP 10 [4] (anti-corrosion pigment)	10.0
Tiona 595 [5] (white pigment)	15.0
Bayferrox 3926 [6] (coloured pigment)	0.3
Finntalc M15 [7] (filler)	20.0
Byk 057 [2] (defoamer)	0.6
Sachtleben micro [8] (filler)	10.0
Xylene (solvent)	0.6
2-propanol/MIBK (50 : 50 solvent mixture)	4.5
Total	**100.0**

Component B	Parts by weight
Aradur 450 [1] (polyamidoamine product)	12.4
Shellsol A [9] (solvent)	6.3
Total	**18.7**

[1] Huntsman Advanced Materials [2] Byk/Altana [3] Troy Chemie [4] Heubach [5] Cristal [6] Lanxess
[7] Mondo Minerals [8] Sachtleben [9] Shell

Guide formulation for water-based 2-component PUR single-layer topcoat

Component A	**Parts by weight**
Tego Dispers 750 W [1] (wetting and dispersing agent)	0.5
Borchigen 12 [2] (wetting and dispersing agent)	0.7
Borchigel LW 44 [2] (rheology control)	0.7
Bentone EW [3] (anti-settling agent)	0.3
Aerosil R 816 [4] (thixotropic agent/filler)	0.5
Texanol (solvent)	0.5
Dowanol DPnB [5] (solvent)	0.5
Dow Corning 6 additive [6] (defoamer)	0.1
Baysilone coating additive 3738 [2] (surface smoothness)	0.3
Tiona RCL-628 [7] (white pigment)	22.0
Westmin 8 E [8] (filler)	7.0
Irgacor 1930 [9] (corrosion inhibitor)	0.8
Heucophos ZCPP [10] (anti-corrosion pigment)	6.0
Butyl diglycol (solvent)	1.0
Plusaqua V 667. 42% WA [11] (polyester-acrylic resin binder)	40.0
Dow Corning 67 additive [6] (substrate wetting)	0.4
DMEA (pH regulator)	0.2
Demineralised water	18.5
Total	**100.0**

Component B	**Parts by weight**
Desmodur N 3600 [12] (isocyanate hardener)	6.0
Bayhydur VPLS 2319 [12] (isocyanate hardener)	6.0
Arcosolv PMA (solvent)	3.0
Total	**15.0**

[1] Tego [2] Borchers [3] NL Industrial Chemicals [4] Evonik Industries
[5] Dow Chemical [6] Dow Corning [7] Cristal [8] Mondo Minerals
[9] BASF Schweiz [10] Heubach [11] Maeder Resins [12] Covestro

Guide formulation for façade paint

Component	Parts by weight
Water	7.0
Coatex P 50 [1] (wetting and dispersing agent)	0.3
Calgon N. 10 % [2] (wetting and dispersing agent)	1.5
Ammonia, 24 % (pH regulator)	0.2
Acticide AS [3] (preservative)	0.1
Tylose MH 4000 [4] (thickener)	10.0
Agitan 315 [5] (defoamer)	0.1
Sipernat 820 A [6] (filler)	2.0
Kronos 2190 [7] (white pigment)	18.0
Talcum Naintsch ASE 10 [8] (filler)	5.0
Omyacarb 5 – GU [9] (filler)	22.0
Butyl diglycol acetate (solvent)	0.7
Dowanol DPnB [10] (solvent)	0.9
Acronal 290 D [11] (styrene/acrylic emulsion)	32.0
Total	**100.0**

1) Coatex 2) BK Guilini 3) Thor Chemie 4) Shin Etsu 5) Münzing Chemie 6) Degussa
7) Kronos 8) Imerys Talc 9) Omya 10) Dow Chemical 11) BASF

Guide formulation for highly filled interior emulsion paint

Component	Parts by weight
De-ionised water	312.2
Coatex P 50 [1] (wetting and dispersing agent)	2.8
Calgon N new [2] (wetting and dispersing agent)	0.9
Tylose MH 30 000 YG8 [3] (thickener)	2.8
Byk 032 [4] (defoamer)	2.8
Mergal K 15 [5] (preservative)	1.9
Tiona 595 [6] (white pigment)	56.6
Omyacoat 420 – OG [7] (filler)	141.5
Omyacarb 5 – GU [7] (filler)	235.8
Omya-Calcimatt – AV [7] (filler)	84.9
Finntalc M 20SL [8] (filler)	47.2
Sodium hydroxide solution, 10 % (pH regulator)	1.9
Mowilith LDM 1871 approx. 53 % [9] (vinyl acetate/ethylene emulsion)	108.5
Total	**100.0**

[1] Coatex [2] BK Guilini [3] Shin Etsu [4] Byk [5] Troy Chemie
[6] Cristal [7] Omya [8] Mondo Minerals [9] Celanese

Guide formulation for high-quality interior emulsion paint

Component	**Parts by weight**
Water	282.0
Tylose H 30000 yp2 [1] (thickener)	3.0
AMP 90 [2] (pH regulator)	1.0
Borchigel VP 97105 [3] (thickener)	4.0
Calgon N new [4] (wetting and dispersing agent)	1.0
Lopon 895 [4] (wetting agent)	3.0
Agitan 731 [5] (defoamer)	2.0
Mergal K 15 [6] (preservative)	2.0
Kronos 2044 [7] (white pigment)	150.0
Finntalc M 30 SL [8] (filler)	90.0
Omyacarb Extra-CL [9] (filler)	120.0
Optiwhite [10] (filler)	30.0
Omyacarb 5-GU [9] (filler)	60.0
Omya-Calcimatt – AV [9] (filler)	100.0
Mowilith LDM 1871 approx. 53 % [11] (vinyl acetate/ethylene emulsion)	150.0
Agitan 282 [5] (defoamer)	2.0
Total	**100.0**

[1] Shin Etsu [2] Angus [3] Borchers [4] BK Guilini [5] Münzing
[6] Troy Chemicals [7] Kronos [8] Mondo Minerals [9] Omya [10] Burgess Pigment

Guide formulation for water-based satin gloss decorative paint

Component	**Parts by weight**
Additol VXW 6208 [1] (dispersing agent)	1.0
Texanol (solvent)	0.5
Dowanol DPnB [2] (solvent)	1.0
Tego Foamex 810 [3] (defoamer)	0.2
Byk 019 [4] (defoamer)	0.2
Rheolate 278 [5] (thickener)	0.7
Rheolate 255 [5] (thickener)	0.2
Bentone EW [5] (anti-settling agent)	0.5
Tiona 595 [6] (white pigment)	17.0
Westmin 8 E [7] (filler)	8.0
Omyacarb 2 – GU [8] (filler)	7.0
Plusaqua P 723, 40 % WA [9] (alkyd resin emulsion)	49.0
Ammonia. 25 % (pH regulator)	0.1
Additol XL 297/100 [1] (anti-skin agent)	0.5
Byk 346 [4] (substrate wetting)	0.3
Nuodex Combi LS [9] (drier)	0.6
Demineralised water	13.2
Total	**100.0**

[1] UCB [2] Dow Chemical [3] Tego Chemie [4] Byk/Altana [5] NL Industrial Chemical
[6] Cristal [7] Mondo Minerals [8] Omya [9] Maeder Resins

Guide formulation for matt paint

Component	**Parts by weight**
Water	19.7
Butyl diglycol (solvent)	0.5
Coadis 123 K [1)] (wetting agent)	1.0
AMP-90 [2)] (pH regulator)	0.2
Dowanol DPnB (solvent)	0.5
Tylose MH 2000 YP2 [3)] (thickener)	0.2
Coapur 2025 [1)] (thickener)	0.4
Byk 038 [4)] (defoamer)	0.5
Alusil ET [5)] (filler)	2.0
Tiona 595 [6)] (white pigment)	20.0
Finntalc M30 SL [7)] (filler)	7.0
Omya-Calcimatt – AV [8)] (filler)	8.0
Omyabrite 1300 X – OM [8)] (opacifying filler)	10.0
Omyacarb Extra – GU [8)] (filler)	6.0
Plusaqua P 755 [9)] (modified alkyd binder)	24.0
Total	**100.0**

1) Coatex 2) Angus 3) Shin Etsu 4) Byk/Altana 5) Ineos
6) Cristal 7) Mondo Minerals 8) Omya 9) Maeder Resins

Guide formulation for powder coating

Component	Parts by weight
Crylcoat 7630 [1] (polyester binder)	674.7
Primid XL-552 [2] (hardener)	35.5
Benzoin (degassing agent)	2.1
Resiflow PV 88 [3] (flow control agent)	8.2
Kronos 2160 [4] (white pigment)	206.5
Sicotan L 2010 [5] (yellow pigment)	1.0
Bayferrox 306 [6] (black pigment)	0.3
Calcigloss – GU [7] (filler)	71.7
Total	**100.0**

[1] Allnex [2] EMS Chemie [3] Worlée Chemie [4] Kronos
[5] BASF [6] Lanxess [7] Omya

Guide formulation for road marking paint

Component	Parts by weight
Fastrack-53, 51.6 % [1] (polymethylacrylate emulsion)	366.0
Nopco 8034 [2] (defoamer)	2.7
Triton X-405 [1] (wetting and dispersing agent)	2.9
AS-238 [3] (wetting and dispersing agent)	8.2
Tiona 595 [4] (white pigment)	96.3
Omyacarb 10 – AV [5] (filler)	456.0
Ethanol (solvent)	11.8
Texanol (coalescence agent)	38.0
De-ionised water	38.0
Total	**100.0**

[1] Dow Chemical [2] Henkel [3] CFCP [4] Cristal [5] Omya

List of filler examples

Filler	Particle size[1] [µm] D50%	D98%	Oil absorption [g/100 g]	Trade name
GCC A	2.2	17	17	Violet Label
GCC B	1.0	4	19	Omyalite 90
GCC C	2.6	10	18	Omyacarb 2 – BE
GCC D	4.0	17	16	Omyacarb 5 – BE
GCC E	6.7[2]	31[2]	15	Omyacarb 10 – BE
GCC F	2.9	13	17	Omyacarb 2 – AV
GCC G	4.8[2]	25[2]	15	Omyacarb 5 – AV
GCC H	11.3[2]	38[2]	14	Omyacarb 10 – AV
GCC I	0.8	4	21	Calcigloss – GU
GCC J	20.0[2]	50[2]	12	Omya-Calcimatt – AV
GCC K	0.8	5	19	Omyacoat 850 – OG
GCC L	0.9	4.5	24	Omyacoat 400X – OM
Dolomite A	5.6[2]	18[2]	13	Microdol 1
Dolomite B	3.8[2]	19[2]	14	Microdol 5
Dolomite C	2.3	12	16	Microdol Extra
Quartz A	1.6	7	25	Sikron SF 800
Quartz B	9.1[2]	34[2]	19	Sikron SH 300
Quartz C	3.4	13	23	Sikron SF 500
Cristobalite A	2.6	10	30	Sibelite M 6000
Cristobalite B	13.2[2]	45[2]	22	Sibelite M 3000
Kieselguhr A	15.5[2]	40[2]	67	Celite 281
Kieselguhr B	6.0[2]	22[2]	125	Celite 499
Talcum A	2.0	10	42	Finntalc M05 SL
Talcum B	8.6[2]	29[2]	31	Finntalc M30 SL
Talcum C	23.4[2]	67[2]	25	Finntalc M50
Talcum D	4.3	20	32	Luzenac 20M0
Talcum E	4.3	17	34	Luzenac 20M00S
Talcum F	8.9[2]	28[2]	24	Westmin D30E

Filler	Particle size[(1)] [µm] D50%	D98%	Oil absorption [g/100 g]	Trade name
Talcum G	2.5	10	41	Micro-Talc A.T. Extra
Talcum H	21.4[(2)]	56[(2)]	21	Steamat
Kaolin A	5.8	24	37	China Clay B
Kaolin B	4.2	23	31	ASP 400
Kaolin C	2.0	13	48	Pole Star 200P
Kaolin D	1.6	13	47	Iceberg
Kaolin E	0.76	6	78	Satintone 5HB
Kaolin F	1.0	10	67	Optiwhite MX
Mica A	2.9	10	37	Aspolit F 30
Mica B	9.5[(2)]	34[(2)]	21	Aspolit F 100
Mica C	5.5	18	28	Plastorit micro
Mica D	8.0[(2)]	28[(2)]	26	Plastorit 0000
Feldspar A	4.6	22	21	Minex 7
Feldspar B	10.8[(2)]	48[(2)]	18	Minex 3
Barium sulphate A	3.2	15	10	EWO
Barium sulphate B	6.0	30	12	Barita Primera 25
PCC A	0.3[(3)]	-	26	Socal P2E
PCC B	0.07[(3)]	-	27	Socal U1S2
PCC C	1.1	-	41	Precarb 100
MCC A	2.4	9	50	Omyabrite 1300 X – OM
Precipitated barium sulphate A	0.9	3	13	Sachtleben micro
Precipitated barium sulphate B	4.0	12	12	Blanc Fixe G
Pyrogenous silicic acid A	0.016[(4)]	-	-	Aerosil R 972
Pyrogenous silicic acid B	0.007[(4)]	-	-	Aerosil 380
Pyrogenous silicic acid C	0.040[(4)]	-	-	Aerosil OX 50
Precipitated silicic acid A	3[(5)]	-	220	Acematt OK 412
Precipitated silicic acid B	4[(5)]	-	180	Acematt HK 125
Precipitated aluminium silicate A	1.2	10	130	Sipernat P 820
Precipitated aluminium silicate B	1.0	7	90	Zeolex 98

Filler	Particle size[(1)] [µm] D50%	D98%	Oil absorption [g/100 g]	Trade name
Precipitated aluminium silicate C	25[(2)]	64[(2)]	112	Optimat 2550
Aluminium hydroxide A	0.8	4	45	Martinal OL-107
Aluminium hydroxide B	10.9[(2)]	30[(2)]	26	Martinal ON-310
Organic filler	0.4	-	-	Ropaque 3000

[(1)] Particle size determined by Sedigraph 5100 unless stated otherwise, [(2)] Particle size determined by Cilas 920
[(3)] Particle size determined by air permeability according to NF X11-683
[(4)] Median primary particle size determined by TEM
[(5)] Median aggregate/agglomerate size determined by TEM, [(6)] Median primary particle size determined by SEM

Author

Detlef Gysau, Omya International AG, has been engaged in the development of hydro-fillers for the automotive industry at Akzo Coatings in Stuttgart, after his apprenticeship of a paint laboratory assistant (1985–88). During his studies he joined the industrial research centre of Rohm and Haas, Philadelphia and the R+D lab for photo initiators at Ciba-Geigy, Basel. 1996 he acquired his engineer degree in the fields of paints, lacquers and plastics (M.Eng.) at the University of Applied Science in Stuttgart. Today, he is employed at Omya International and led for 14 years the Applied Technology Services for Paints, Coatings & Adhesives (ATS-PCA) with global responsibility for development and technical service. In 2010 Detlef Gysau finished his Executive MBA in General Management at the University of St. Gallen, Switzerland and changed to Omya's Group Function Sales & Marketing. Here he held positions as Marketing Strategy Manager, Head of Product Management and actually Head of Innovation & Technical Marketing in the Segment Construction.

Index

Symbols

A

B

C

Detlef Gysau: Fillers for Paints

D

E

P

Q

R

S

T

U

V

W

X

Y

www.ingramcontent.com/pod-product-compliance
Lightning Source LLC
LaVergne TN
LVHW021726021026
849161LV00004BA/750

* 9 7 8 3 8 6 6 3 0 2 8 4 6 *